Principles of Security Management

ᴇʀ BOOKS BY THE AUTHORS

ᴄhard J. Healy and Dr. Timothy J. Walsh —

"Protection of Assets Manual." Santa Monica, California; The Merritt Company.

"Protecting Your Business Against Espionage." New York: American Management Association.

Richard J. Healy —

"Design for Security." Second Revised Edition. New York: John Wiley and Sons, Inc.

Principles of
Security Management

Richard J. Healy
Dr. Timothy J. Walsh

PROFESSIONAL PUBLICATIONS
Post Office Box 698
New Rochelle, NY 10802-0698

1st Printing 1981
2nd Printing 1983
3rd Printing 1985

International standard book number 0-9605954-0-6
Library of Congress catalog card number 81-81449

Contents

Preface
To First Edition

THE chief reason for this book is the conviction that business losses from crime, violence, disorder, and disaster can be controlled.

Despite appalling crime statistics, spreading unrest and militancy, and a marked relaxation in personal standards of conduct, responsibility for the occurrence of major business asset losses must still be charged to the business manager. His ignorance or neglect encourages them; his awareness and action avoid or reduce them. No modern manager, whether of a business for profit or of a public or private institution, can achieve complete success in his task unless he pays appropriate attention to the identification and control of asset losses.

To do the job, a manager needs, first of all, to be aware of its scope and then to possess or have access to the skills, techniques, and resources that are required. It is the purpose of this book to provide a systematic approach to filling both needs. Discussions of the size, trend, and character of security loss risks are balanced by explicit descriptions of the ways in which losses actually occur and, most important, by specific recommendations for countermeasures.

The context in which asset losses should be studied is quite different today than it was a decade ago. Major social forces,

acute political pressures, and fundamental changes in jurisprudence have combined to generate strikingly new and different
attitudes about crime, property, the resort to violence, and the
obligations of the individual citizen. It is one thing to be concerned that finished goods or raw materials may be stolen unless
safeguards exist. It is quite another thing to fear that a bomb
or incendiary device may be used against the enterprise by an
extremist. Security planning today must consider the increased
probability of both risks.

Another factor that has changed security management is technological advance. In the security field, technology also includes
a kind of shadow image that follows just behind the useful developments and reflects new or improved criminal techniques. With
the advent of large-scale computer applications has come large-
scale fraud and theft through use of the computer. With micro-
miniaturization and solid-state communications engineering has
come increased interception of personal and business data for
criminal purposes. Even rather simple advances like the thermal
lance have made years of safe and vault design obsolete. At the
same time, technical advances can be turned to positive use in
programs of security control. The ubiquitous computer, in a quite
new role, now directs the operations of integrated security and
protection systems. Communications techniques of multiplexing
and polling have made wide-area and long-distance alarm and
surveillance coverage economically feasible. In short, the new
technology is both a promise and a threat. The successful enterprise manager must distinguish and respond.

In dealing with security problems and solutions, the function
of this book as a management tool has been kept in prime focus.
Works that pursue special aspects of security in a more comprehensive or technical fashion are available. Reference is made
to them in the text for readers who want to follow a subject
further. Two earlier books by Richard J. Healy are particularly
useful for developing physical security systems and for disaster
control planning; they are *Design for Security* and *Emergency
and Disaster Planning*, published by John Wiley & Sons. These and
other sources referred to make excellent supplements for those
who are in need of more specialized information. The present

work is intended as a first source for any manager with asset protection responsibility.

Strictly, every manager has some asset protection duties, but there are functions in which the responsibility is more concentrated. Financial, purchasing, data processing, insurance, distribution, and security managers are in that category and will find many typical security problems discussed in this book. But research directors and personnel managers also have assets protection concerns. The success of many an enterprise lies in its innovations, discoveries, and improvements. They will involve trade secrets and proprietary information, and the research director is responsible for protecting them.

The vulnerability of any organization to security losses caused by its own employees is measured in part by the character of the employees, their awareness of assets protection requirements, and their motivation to cooperate. The personnel manager is concerned with the selection and training of employees from the security point of view as well as from other viewpoints. Both the research director and the personnel manager will find here material relevant to their concerns.

Many colleges and universities are adding security programs to or expanding them in their undergraduate and graduate curricula. Students of business can now find opportunities for formal education in assets protection as well as in the more familiar management disciplines. To provide perspective for the application of such theoretical security studies to the concrete problems of management is one other objective of this book.

Some concepts and practices discussed in this book will be familiar. In the experience and judgment of the authors they are still valid, useful, and appropriate. Other ideas presented here will be new to some readers. Nothing offered is untried, however. All the approaches and techniques have been used by the authors and have passed the crucial test: they have worked. They are presented in the confidence that they will do so again.

Richard J. Healy
Timothy J. Walsh

Preface
To Second Edition

THE first edition of this book, which was published by the American Management Association in 1971, was prompted by the conviction that business losses from pure risk causes could be controlled.

Continued participation in loss control activities and continued observation of business community reactions to loss threats have strengthened that conviction and also sparked some considerations which have led to this new edition. Among them is the idea that not only can losses be controlled but that the direct economic benefit of a properly conducted security program can be shown objectively. This is of central importance to any organization which requires its resources investment to show positive return.

Another reason this revision was undertaken was to take specific cognizance of problems in the security field which were only present in outline form when the work was first done. Terrorism is a prime example of these problems and a substantial treatment of that topic has been added.

Aware, also, that new practitioners in the field are coming increasingly from community and four year colleges and from graduate programs in security and criminal justice at universities, the work has been modified by the addition to each chapter of review and self test questions. These will help make the book more useful in formal study situations.

To keep costs as low as possible, the format has changed to soft cover with little reduction in its expected shelf life.

With these additions and changes, the work is still addressed to its primary audience—working security professionals, managers charged with asset protection duties, and students preparing for a career in security or general business management.

Richard J. Healy
Timothy J. Walsh

= 1 =

Security and Assets Protection

NO enterprise, profit or non-profit, can operate and achieve its purpose without assets. By broadest definition, an asset is "an item of value".[1] In somewhat more restrictive accounting usage, an asset is "anything of value owned."[2] The idea is clear—an asset is something owned by someone for whom it has value.

A business entity, irrespective of whether it be a simple proprietorship or a complex corporate venture, can be defined as the mixture of assets and liabilities whose use or activity constitute the life of the business. In the same way, any organized activity, charitable, public service or whatever, can be said to be constituted of assets and liabilities and to function specifically in the way those particular assets and liabilities are used.

It is quite evident, then, that for an enterprise to continue it must hold and conserve its assets. Loss or reduction in value of those assets necessarily tends towards failure of the enterprise. In the normal course of events each venture will place its assets at risk in some way. For the profit making business the conventional risk will be to use the assets (materials, tools, equipment, land and all else of value) in such a way as to create a tangible product or provide a measureable service that was not available before or was not available in the same

NOTE: References appear at the end of each chapter.

1

way before. That product or service is then offered for sale in the hope that after a period of time the income derived from the sales will not only replace the asset values which were consumed in producing the product or rendering the service, but will provide additional value (profit). The maintenance and replacement of the original asset value is essential to continued existence. The provision of profit is essential to growth and economic success.

The not-for-profit venture operates in the same way with the exception that it does not seek "profit" as added value. It does seek to replace or maintain the original asset value and often to create a "surplus" which is used to expand the venture so it can serve a larger or broader purpose.

Whether a venture is a business for profit or some public service or non profit activity, it absolutely must conserve and replace its consumed assets.

But the exposure of assets to risk in a conventional way, as through normal competitive business operations or usual public service activities, is not the only manner in which those assets can be exposed. Extraordinary exposure—the vulnerability of assets to loss in ways not considered within the scope of normal activities—can result in the loss of those assets and the end of the life of the enterprise. It is in measuring, reducing and controlling the extraordinary exposures that Security makes its most important contribution.

Security Risk Classification

The proper beginning is a security vulnerabilities analysis or classification and arrangement of loss events that will fall within the boundaries of the security program. Not every business loss will be a security loss. For example, the loss of sales due to effective, but legal, competition is not a security loss. The loss of key manpower to another employer who offers a better inducement is not a security loss. Complete loss of a market because a legitimate competitor introduces a product sooner is not a security loss. Each of these losses could have been a gain. In each case there is the contingent possibility that the enterprise will perform successfully by increasing sales, hiring the key recruit, or penetrating the market. Those situations are related to profit and loss through competitive operations in the marketplace. Whatever the function of a business line manager, his daily activity includes decisions and choices intended to optimize profitability. But there are loss situations that do not fit that model, and they are the subject of security consideration.

Losses that do not have the offsetting possibility of gain—pure losses, as the insurance underwriter would call them—involve risks and threats that the average manager may never have had to analyze before. The following list is a summary, by category, of vulnerabilities that are common to many enterprises. The list does not exhaust security vulnerabilities, but it does include all the primary risks and all the risks that are found throughout all industry. A security program that considers those risks will be a major step in planning; and if the proper countermeasures are selected, it will offer a very high degree of reliability that what should have been done to prevent and control security losses has been done.

Some Common Security Vulnerabilities

War or Nuclear Attack

Natural Catastrophe

Tornado	Hurricane
Earthquake	Flood

Industrial Disaster

Explosion	Fire
Structural collapse	Radiation incident
Major accident	

Sabotage and Malicious Destruction

Incendiary fire	Bombs and bomb threats
Labor violence	Civil disturbance
Vandalism	**Terrorism**

Theft of Assets

Pilferage	Embezzlement
Fraud	Industrial espionage
Records manipulation	Shoplifting
Forgery	Robbery and hijacking
Car thefts	

Conflicts of Interests
Employees with their own businesses
Employees working for competitors
Kickback situations

Personnel Problems

Gambling	Absenteeism
Loansharking	Misrepresentation
Disaffection	Narcotics
Disturbed persons	Antisocial behavior

Miscellaneous Risks

Traffic accidents	National security problems

The arrangement of the selected vulnerabilities in the preceding list is keyed to "criticality," a term discussed in detail a little further on. In essence, criticality is the measure of impact of a security loss. How much will it really hurt if the loss event occurs? Remember that some of these vulnerabilities high on the criticality scale are showing a marked uptrend in the United States and the world. Incendiary fires and bombings, in particular, have increased in recent years, and the public statements of various militant groups tell us that terrorist techniques involving those loss risks are a regular part of strategic policy and tactical planning. Until the forces of conflict are quieted and tensions in political, industrial, and social relations are eased, it is likely that losses caused through violence will increase.

But even if extraordinary losses through violence were not a growing menace, there would be little reason to remain indifferent to other types of security losses. Crime in the United States rose 176 percent over the decade of the sixties and continued to rise through the 1970's. The FBI Uniform Crime Reports—the sources of crime frequency data—are criticized by some because they are affected by such factors as improvements in the reporting of crime. Assuming there are fluctuations in crime statistics that are not caused by actual changes in the types or number of crimes committed, it should still be evident that crime itself is dramatically increasing. In the 1967 report of the President's Commission on Law Enforcement and the Administration of Justice, survey data about crime frequency were presented independently of the Uniform Crime Reports. According to the report, the surveys showed the actual amount of crime in the United States to be several times that reported in the UCR.[3] The same report also noted,[4]

> Property crime rates . . . are up much more sharply than the crimes of violence. The rate for larceny of $50 and over has shown the greatest increase of all Index offenses. It is up more than 550 percent over 1933. [1933 was the inaugural year for Uniform Crime Reports data.]

In a recent report, the U.S. Department of Commerce made available data on crimes against business. In the foreword to its findings the report says that crime against business cost $23.6 billion in 1975.[5] In an even later study the American Management Association estimated the annual cost of business crime to be $40 billion.[6]

Serious though they are, the loss vulnerabilities from crime and violence are of less concern than those from catastrophe and disaster. Fires, which now exceed $3 billion annually in the United States, floods, and major industrial accidents are a much bigger potential because of the dimensions of any single loss. As is well known, two-thirds of small businesses affected by serious fires do not reopen. Major industrial and commercial concerns are better able to absorb serious fire losses and not go out of business, but they too are finding increasing difficulty because of the refusal of casualty insurance companies to issue or renew policies. Recent unfavorable underwriting experiences by the carriers mean that even the large firms will be feeling the impact of catastrophic losses much more intensely than in the past.

Aggravating the catastrophe problem is the regular growth in the size and aggregate value of assets exposed to single-incident losses such as a major fire or flood. The dependence of almost all major businesses on the daily output of computers and electronic data processing facilities has made such equipment a prime loss hazard not only because of the value of the physical assets but because of the vastly larger related assets whose value cannot be realized without continuing computer output. Loss of a computer through accident or disaster may have staggering cost consequences because of delay or termination of functions that depend for their control upon the computer output. One example will illustrate. Any transportation organization that uses computer calculations to reserve or assign space (airlines, railroads, over-the-road carriers) will not be able to optimize space utilization if the computer is unavailable. The revenue loss could amount to as much as $1 million a day for large international carriers.

Recognition of Security Vulnerabilities

After the manager has placed general assessment of security hazards in the suggested context, he is ready for the next step in the process: the identification of specific vulnerabilities.

Recognition of a specific security risk often requires specialized skill; it is not entirely a commonsense activity that can be reliably done on a self-help basis. For example, opportunities to gain access to an industrial plant or business office do not consist only of entry through doors and windows. Access can be gained to premises through openings such as utility service entrances, ventilators, and roof hatches or even by the forcible penetration of building walls. Awareness of these unusual access opportunities is essential to a workable security plan. A supply of cash or of items of intrinsically high value will attract theft attempts irrespective or whether the thief must enter through a door or an air conditioning duct or a hole in the wall. If all options are available to him, the thief will most probably select the one that requires least effort and involves the lowest likelihood of detection. However, denying him one or two and leaving a third may not deter him; it will only reduce his alternatives. If the option left is safe and easy enough in his judgment, the potential thief will take it.

The elimination of some security loss vulnerabilities while others are ignored or discounted may be a more expensive approach than having done nothing, because the loss may still occur and have to be borne in addition to the costs of the ineffective security precautions. The ability to recognize real vulnerabilities is primary in the assessment of loss potential.

Each type of loss has its own complex of circumstances and conditions that make its occurrence possible. It is not usual for a business manager or owner to be personally familiar with all of the factors involved in each of the vulnerabilities present within his enterprise. However, there are specialists whose function it is to be familiar with them. The industrial security manager or safety and security director or risk management supervisor, as he is variously called, has the prime task of identifying and planning for the control of all or certain classes of security vulnerabilities. For organizations that do not have such staff capability, consulting organizations comprised of personnel with the skill and experience for the task are available. Whether the resource is internal or external to the business, the important point is that it be present. A vulnerabilities analysis performed by persons not expert in security risk appraisal will almost certainly lead to improper countermeasures decisions.

When done properly, a vulnerabilities analysis is comprehensive and accurate and leads directly to appropriate countermeasures. As circumstances and conditions change materially, risks must be reappraised. The vulnerabilities assessment is not a task done once; it is a continuing effort.

The Use of Matrix Techniques

In analyzing the extent of particular vulnerabilities, the use of grids or matrixes will assure that all relevant factors are considered in establishing the possibility of the loss. They will also contribute a great deal toward the further assessment of the degree of loss probability that should be assigned. Table 1 is a typical matrix used for analysis of the vulnerability to theft of cash. It lists all the places at which cash is maintained on the premises, and for each place it lists the major factors that could lead to cash theft. Physical situation, accountability records, alarm protection, cash storage facilities, control and surveillance of admittance, and the presence of bait money are important among such factors. Also important is the known history of cash losses from the place in the past. Because most business enterprises do not operate on a 24-hour basis, it is necessary to distinguish between precautions during working hours and those during nonworking hours.

The illustrative matrix could have been extended to include many other relevant but somewhat more remote factors. For example, it would be relevant whether the cash was in coin or paper. Large amounts of coin are not nearly as attractive a theft target as smaller amounts in paper. The selection of precisely which factors are to be included in the matrix is a judgment that requires the specialized skills mentioned earlier. For each security risk identified at the particular location, an appropriate matrix should be developed. Careful examination of the facility will reveal the places at which the threat could be actualized. Such vulnerabilities as flood and hurricane cannot readily be localized in that fashion. For those risks a matrix would be designed to reflect the type of damage or consequence on a localized basis.

The matrix serves as a summary and as a trend marker. In one place and in an integrated fashion the risk factors pertaining to a particular loss vulnerability can be arranged to permit rapid review. By noting whether the same factor shows up in many places, it is easier to mark general weaknesses. Thus, if in an extended analysis of many vulnerabilities within a facility it appeared that risk of loss due to theft or risk of unauthorized admittance was frequently increased because of the absence of locking devices, a general security weakness requiring systematic attention would have been noted.

The most important function of the matrix is its value in the assignment of probability. A possible security loss becomes a greater problem as the probability of the occurrence increases. Of fundamental importance in designing security countermeasures is the quantification, to the maximum possible extent, of the probability of occurrence.

Loss Probability

If two security loss vulnerabilities exist, of approximately equal consequences in severity, but one is of a higher order of

Table 1. Cash Theft Vulnerability Matrix

Building Location	Amount on Hand, Dollars		Accountability Records		Area Has Physical Bounds		Area Locked		Positive Control on Admittance		Alarm Protection		Surveillance Devices		Cash in Storage Container		Bait Money Kept		History of Cash Loss	
	NBH	OT	NBH	OT	NBH	OT	NBH	OT	NBH	OT	NBH	OT	NBH	OT	NBH	OT	NBH	OT	NBH	OT
Cashier's office	20,000	5,000	Y	Y	Y	Y	N	Y	N	N	Y	Y	N	N	Y	Y	Y	Y	N	N
Manager's secretary	300	300	Y	N	N	N	N	N	Y	N	N	N	N	N	Y	Y	N	N	Y	Y
Cafeteria	1,000	0	N	N	Y	Y	N	N	N	N	N	N	N	N	Y	—	N	—	Y	N
Employees' store	500	0	Y	N	Y	N	Y	Y	N	N	N	N	N	N	Y	—	N	—	N	N
Reception	100	0	N	N	N	N	N	N	N	N	N	N	N	N	N	—	N	—	N	N
Etc.	Etc.	Etc.																		

NBH = normal business hours; *OT* = other times; *Y* = yes; *N* = no.

probability, with which should the manager concern himself first? The answer is clear. The more likely event is the more troublesome. But there will not be one case in which two vulnerabilities compete for attention. In even the most modest enterprise there will be dozens, and in the major industrial installations there will be thousands. If it is obvious that priority is indicated when only two vulnerabilities exist, it follows that priority is even more necessary when dozens or thousands of risks must be considered. Without a systematic theoretical basis on which to proceed, time and money, always limited, may be expended on security risks of lesser importance while serious problems wait. That is exactly what does happen in the enterprises that do not require the security function to be accomplished against objectives-oriented criteria of the kind applied to other management assignments.

But probability can be measured with mathematical precision in some situations. Is it possible to reduce security risk probability measurement to precise formulas? Interestingly, a 1967 Small Business Administration report[7] contains an appendix entitled "Protective Devices Systems," developed by Stanford Research Institute, that suggests a formula for equating losses at protected business sites to losses at unprotected sites. The formula is factored for various sets of probabilities concerning entry, intrusion detectors, and physical security measures.

Underlying data for the Stanford study were taken from Underwriters Laboratories field service reports from 1962 through 1967. The reports summarize actual loss situations according to the protected or unprotected condition of the premises, the methods used for entry, and other features; and create a measured population for deriving the probability of occurrence. The formula provides general guidance to small business in the construction of a cost-benefit model for determining a reasonable dollar investment in physical security. But large commercial or industrial complexes are different from small businesses. Many security vulnerabilities are partly neutralized so that losses do not occur to be reported. Other losses are absorbed through techniques of risk assumption and are not reported anywhere outside the enterprise. Until there is a larger, reliable collection of base data concerning industrial security vulnerability, it is better to establish probabilities on some basis that, though lacking in mathematical precision, will be oriented toward the conditions peculiar to larger commercial and industrial operations.

One approach to probability determination that is fairly easy to apply involves a rough grading technique in which levels of probability are broadly separated by a partly subjective evaluation. An event's probability can be included within one of five statements: (1) the event is virtually certain to occur; (2) the event is highly probable; (3) the event is moderately probable, that is, equally likely to occur or not to occur; (4) the event is improbable; or (5) the probability of occurrence is unknown. For our purposes "certain" here means physically certain. For example, if two equidistant speeding automobiles traveling at the same rate are approaching a blind intersection with very narrow swerving or turning room, it is physically certain that they will collide. Similarly, if a fire starts in an unattended flammable fuel storage shed that is not equipped with detection or extinguishment facilities, it is certain that there will be an explosion or conflagration or both.

By using the matrix technique and cumulating the factors favorable to the occurrence of a security loss event, it should be possible to assign one of the first four levels of probability. There will, of course, be those situations in which it cannot be said whether the event falls clearly within one level of probability or within the next higher or next lower level. Then a rule of thumb assigns the higher of the two levels to assure against underestimating.

A practiced security specialist should not have difficulty in distinguishing most vulnerabilities if the matrix is adequate. However, there will be cases in which factors are not fully developed or historical information has not been collected. The probability-unknown category is then appropriate. The vulnerabilities about which too little is known for a sound judgment are noted for later completion of the data gathering. No complete security vulnerability analysis should retain any vulnerability in the probability unknown class. That is a provisional rating to be replaced with a firm evaluation after further investigation.

To be sure that security vulnerabilities have been observed

wherever they exist, it is important that every process, every operation, and every location in the facility be scrutinized. In all instances the search will be for factors that can be shown to favor occurrence of the loss. Those factors will fall into one of the following classes: (1) *physical environment,* including location, composition, and spatial relationships, (2) *procedural aspects,* (3) *history,* particularly history of past losses, and (4) *criminal state of art.*

The fourth class deserves comment. To know whether a security vulnerability involves high or low loss probability sometimes demands a technical awareness of the range and capability of tools available to an aggressor. Safes and vaults, for example, provided an entirely different level of physical security against forced entry before the development of the thermal lance than they have since. Microminiature radio transceivers featuring integrated circuits can be concealed in places not practicable before and can operate longer on low-output power sources.

The impact of technical developments on vulnerability to industrial espionage is of primary importance. To the manager who does not appreciate the difference between a latch, a dead bolt, and a deadlocking latch bolt, it may be a matter of passing interest which device secures his plant after hours. However, to the attacker seeking entry, the difference may determine whether he even makes an attempt at entry. As each security device or procedure is introduced, it is merely a matter of time before a tool or technique is found to defeat or circumvent it. Knowing the array of the tools and techniques is a necessary part of the assessment of vulnerability.

Probability, important though it is, is not enough for the development of cost-effective countermeasures. An event may be highly probable, even certain to occur, but be of such small consequence in terms of impact as not to deserve attention, or at least not priority of attention. The second characteristic to be evaluated for each vulnerability is its criticality.

Loss Criticality

Most simply defined, criticality is the impact in dollars of cost of a security loss. Dollar cost is the measure because the

ultimate criterion of any business enterprise is dollars. It is sometimes heard from security specialists that they find managements unresponsive or insensitive to security problems. In most
of those cases further study reveals that the management has
not been given a proper basis for response. If an executive with
profit-and-loss responsibility is to determine whether to resource
a security program with available dollars, he must be able to
discount those dollars in alternative cash flow patterns. A security
program should not be funded unless its requirement for dollars
is clearly related to the dollars of loss it will probably prevent.

It is in this area that serious oversights are committed both
by security staff specialists and by general management. Although a security manager or consulting adviser can identify
the types of security vulnerabilities and even assess the probability of their occurrence without direct help from the enterprise,
it is improbable that either will be able to determine the criticality of a loss without assistance. The reason is the need for a
different kind of skill and knowledge: that related to costing.

Criticality, measured in dollars of cost, can involve four
principal types of real costs. The first is permanent replacement
cost. If an asset is lost or destroyed (as by fire or theft, for
example) it may be necessary to replace it to continue the operations of the enterprise. In the case of a stolen machine tool,
or a burned electric transformer, or a damaged computer main
frame, that would be true. The cost of the permanent replacement must be considered the first item in the loss cumulation.
Permanent replacement, depending upon whether the replacement asset is purchased or made by the enterprise, may include
(1) purchase price, (2) freight and shipment, (3) labor, (4)
materials costs, and (5) other costs. One example of other costs
would be travel or communications expense necessary to locate
a source from which to procure the permanent replacement.

The second category of cost is temporary replacement cost.
Not every asset that is lost or destroyed will require a temporary
replacement, but many assets will. If sabotage destroys a piece
of manufacturing equipment, delivery and contract commitments may require that, pending permanent replacement or repair, some other equipment be used. The other equipment may

also be working full time and require premiums of one kind or another to extend its use to pick up the production normally accounted for by the damaged equipment. Or it might be necessary to subcontract the interim work and pay subcontractor's costs and profit. The cost of a temporary substitute should logically be attributed to the security loss that made the usual equipment unavailable.

The third type of cost is related cost; it is typified by idle time or waiting time expense. For example, following a civil disturbance in which incendiary devices are hurled through plant windows and severely damage some delicate process control equipment, processing must stop until the control gear is restored to service. The costs of the stoppage are related to the security incident that damaged the control gear. In any accurate appraisal of the cost of loss, they should be included.

A fourth class of cost involves discounted cash. If the cash that must be diverted to pay some or all of the first three classes of cost had remained available for investment, what would it have returned? Since it isn't available, the lost return should be included in total costs, particularly if the enterprise is otherwise seriously committed to short-term money management.

Some of the costs of loss may be reimbursed or indemnified through insurance. Others may not be. Aside from the question of assumed risk or large deductible provisions that will reduce available indemnification, costs may fall into classes that for legal or policy reasons cannot be indemnified. For example, if a trade secret is improperly disclosed and becomes part of the public domain, the cost of the development program that led to it will not be indemnified. If, in addition to the lost profits through hijacking of whole shipments of a product, market is also lost because of disgruntled distributors and dealers who cannot meet the buying demand, loss of that market will not be indemnified. If, in an attempt to foil a robbery on the premises, an armed guard employed by the enterprise discharges his weapon and mains a bystanding visitor and it develops during the trial of the damages action that the enterprise neither trained the guard in firearms nor made a serious inquiry as to whether he had ever received training elsewhere (a not uncommon situation), there is a high probability that such an enterprise may be found to have been grossly or wantonly negligent. Public policy usually denies indemnification for gross or wanton negligence.

Degree of Criticality

When all the costs reasonably attributable to a security loss have been totaled or approximated, there is a basis for assigning a criticality level. As with the earlier assignment of probability, it is suggested that a broad distinction be used. If four levels of criticality are employed, roughly parallel to the four levels of probability, we might regard a loss as (A) *fatal* to the business if the cost is so high that the business is terminated, (B) *very serious* if the impact, although not enough to destroy the enterprise is enough to require major adjustment of investment policy, (C) *moderately serious,* if the impact on earnings and return on investment is sufficient to require some comment to the equity owners, (D) *relatively unimportant,* if the cost can be absorbed within existing contingency reserves or if overall profitability is affected only slightly, (E) *seriousness unknown,* when all the cost factors have not yet been identified or quantified. Like "probability unknown," "seriousness unknown" is a provisional assignment and will be replaced by a definite criticality judgment after necessary research is complete.

After each vulnerability has been assessed as to probability and criticality, it is possible to combine the ratings to arrive at a priority sequence. Figure 1 summarizes the suggested ratings. Each vulnerability will be assigned a combined ranking symbol. For example, a given vulnerability might be rated 1-A. Almost certain to occur and fatal if it does, such a vulnerability would head the list of priorities. Other vulnerabilities might be rated 3-D or 4-C and would take a low place in the counter-measures schedule.

In setting final priorities there are many combinations or sequences that could logically be urged as proper. On balance, it is best to give greater weight to criticality than probability. Countermeasures plans would first consider all losses of moderate or higher probability that involve very serious criticality. Only then should losses involving moderate criticality be considered, even though they are of a very high probability.

Systems Logic

When priorities for countermeasures are set, systems logic should be used to optimize the program design. Optimum security countermeasures programs are those that provide adequate safeguards at an acceptable level of confidence in their reliability for the least cost. All the terms of that definition are important: sufficient countermeasures, adequate reliability, least cost. If there is premature emphasis on or exclusive concern with its cost, the program will not achieve satisfactory results. Cost effectiveness, not absolute lowest cost, is the standard. To achieve

Figure 1. Basis for Assessing Vulnerability

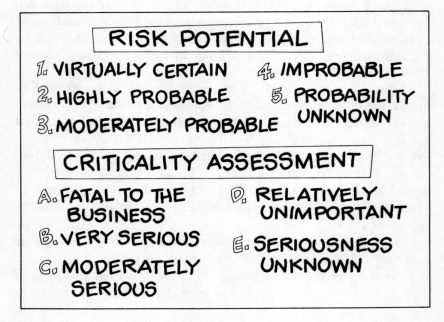

RISK POTENTIAL

1. VIRTUALLY CERTAIN   4. IMPROBABLE
2. HIGHLY PROBABLE   5. PROBABILITY
3. MODERATELY PROBABLE   UNKNOWN

CRITICALITY ASSESSMENT

A. FATAL TO THE   D. RELATIVELY
   BUSINESS   UNIMPORTANT
B. VERY SERIOUS   E. SERIOUSNESS
C. MODERATELY   UNKNOWN
   SERIOUS

**Figure 2. Logical Sketch of the Security Risk
of Undetected, After-hours Entry**

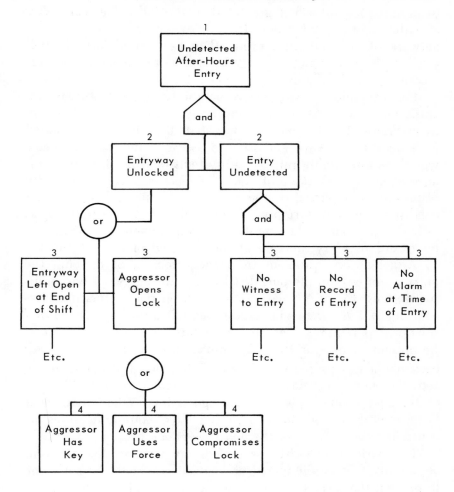

1. Security risk
2. First set of conditions
3. Second set of conditions
4. Third set of conditions
Etc. Continued development of logical structure to identify control points, that is, points at which introduction of a security countermeasure could affect the risk

cost effectiveness, systems logic will be helpful because it takes into account the enhancement of one countermeasure by another to produce a joint effect greater than the individual capability of either alone. Also, the risk complex, regarded as a related network of vulnerabilities rather than as isolated hazards, will suggest places at which leverage is possible to make one countermeasure do several jobs.

For example, consider a vulnerability to surreptitious unauthorized access. If the problem is set out in a block diagram, as in Figure 2, it becomes clear that such access can be gained in a number of ways and that to assure denial of the access would require highly reliable subsystems of control on the issuance of keys, the locking of doors after hours, and the reliability of personnel employed by the enterprise. To neutralize the other condition, that of surreptitious or undetected access, requires only that any one of the simultaneous requirements of no witness, no record of entry, and no alarm or signal of entry be met.

Attacking the problem from the direction of denying the entry requires multiple remedies because the aggressor has multiple options. Every *or* relationship between risk factors indicates that *one or the other* will serve to actualize the risk. Every *and* relationship requires that all the factors connected by the *and* be present. Arranging the risk complex in logic tree format, with each major risk properly broken into the constituent options for actually converting the risk to a loss, will allow identification of the *and* connections, the leverage points. When they have been identified, the analysis of available countermeasures to neutralize one or more of the *and* factors can proceed.

The basic logic of the system design tells us that any one of the *and* factors eliminated or neutralized should be sufficient to prevent that loss. By increasing the number of factors neutralized, a higher reliability can be built into the program. Alternatively, by using several techniques responding to different principles, the assurance that any single factor has been neutralized will be strengthened.

An illustration will help. If detection of fire is important to neutralize a vulnerability to fire loss, then the use of one type of detector to sense critical changes in temperature and another, independent sensor to detect the presence of smoke or combustion products will greatly increase the probability that an actual fire will be detected, because the likelihood of simultaneous failure or malfunction of two detectors, operating independently and on different principles, is much more remote than that of failure of either alone.

Other chapters in this book, particularly Chapter 10, "The Systems Approach to Security," discuss detailed applications of a variety of manual, procedural, electronic, and electromechanical countermeasures in integrated systems. The design and development of such systems is the complement and consequence of systematic analysis of security vulnerabilities. It is important not to lose sight of that relationship.

A variety of security countermeasures can be arranged in some internally consistent way and labeled a "security system." Indeed, there is much promotion today of just that approach to security. Such a system can include highly reliable and technically advanced components. It can even be optimized for some model threats. However, unless the countermeasures system relates directly to the vulnerabilities system, there is the possibility that the end result will not be what is required; in fact, it will usually be more than is required. Ready-made systems will add subsystems or components to the standard configuration whenever an application requires it but will not usually eliminate components that are not required for that problem. It is often the case of chrome plating where black metal would have done. As the cost of security countermeasures programs climbs, it is increasingly important that overdesign or chrome plating be eliminated to assure the greatest return on the invested funds. Systems logic on both counts, threat analysis and countermeasures design, will help to find the risk, match the remedy, and allow for required updating.

The Economics of Security

From the preceding comments it is evident that in economic terms, the security function is to conserve the assets of the enterprise against pure risk loss or value reduction. But Security, itself, will require resourcing. Capital and expense costs will be generated for personnel, equipment and operations. To the extent these costs delay or impede the application of income to asset replacement and profit or surplus they are counter-productive. It is necessary, therefore, to assure that the results of the security program, in economic terms, exceed the costs of the program. Said another way and in orthodox investment language, the security operation must yield a return acceptable to the enterprise.

Although it is not possible (or even desireable) to insist that security results be justified exclusively by return on investment, that yardstick should be used to the maximum possible extent because it is the measure used on all the other enterprise functions. So, it would not be appropriate to argue against an adequate personal protection effort for exposed key employees in world areas with high terrorism potential merely because such a program, even if successful, could not be justified in economic terms. (Although it is an interesting, but puzzling fact that some enterprise leaders do reject such protection efforts, even when they are personally at risk, on economic grounds. It is probably a mixture of unwillingness to appear frightened—the "macho syndrome"—and frustration at being unable to identify clearly what precise measures could be taken which lead to the preposterous conclusion that expenditures to protect threatened key personnel are economically unproductive.)

On the other hand, it would be doubtful management wisdom to continue a corporate or divisional security program costing substantial amounts annually without some indications that there was a proportionate return. If it cannot be shown, even inferentially, that a $500,000 annual investment in a security program is returning at least the $500,000, then that sum could be invested in certificates of deposit at 14 or 15% annually. In the first case the result would be a reduction of enterprise cash (an asset) of $500,000 with no measureable return. In the latter the result would be $570,000 or $575,000 cash at year end. There must be a tangible benefit resulting from the diversion of a half million dollars and the loss of $70,000 or $75,000 income to satisfy a reasonable manager. Unlike conventional law enforcement expense which communities accept because to be without police would expose the public to a risk no one is willing to assume, the cost of security is not approved by general concensus. Nor ought it to be. Allocation of all resources within a well run business or public service enterprise should be competitive and made on the basis of the measureable extent to which the recipient activity contributes to the prime purpose of the enterprise.

Measurement Techniques

The costs of the security program can be determined by standard techniques used by all enterprises. Personnel expense will consist of salaries, benefits, recruitment and training elements. Other expenses will include consummable supplies, travel, rentals and purchased services. Capital costs will include allocable portions of depreciation on capital assets used in the security effort. In regard to the capital items it is also important that they be approved initially using some form of payback and discounted cashflow analysis, i.e., that the foreseeable economic benefit be measured in advance, and then that the delivery of that benefit be continuously monitored as suggested in the next few paragraphs.

The cost of security losses is also a vital part of the measure and can be seen in two perspectives. First is the avoided cost of loss. Using probability estimates based upon the relative probabilities described earlier and factoring them by the estimated criticalities of identified losses, it is possible to derive some numerical estimate of the probable cost of a security loss. For example, if it is agreed, given no change in situation, that a particular larceny loss is a "highly probable" occurrence, or that a specific disaster, say a flood, has a historic frequency of occurrence of one every three years, then impact of those occurrences could be calculated as follows. For the larceny, a probability factor of .75 could be used to represent "highly probable". If the amount at risk was known to average $75,000, then the impact of that loss could be stated as $75,000 × .75, or $56,250. The last number would be weighted for the amount at risk and the estimated probability of a single occurrence during the time period being used (usually a calendar or fiscal year). If the larceny would be as likely to be committed many times as once and if the amount at risk, even after the first larceny, would remain at approximately the same level, the loss impact would be increased by some multiple to allow for repeated thefts.

In the case of the flood, because we know it occurs at a frequency rate of once every three years, it is possible to assign to any one year a proportionate share of the cost. If it can be estimated (or is historically known) that a flood will generate costs of $1,500,000, the impact of floods, given no change in the nature of the exposure, can be calculated as $\frac{\$1,500,000}{3}$ or $500,000 annually. Obviously, if the flood does occur it will generate all the costs in the year of occurrence, not one third the cost. But, for planning and resources allocation it is appropriate to spread the cost of occurrence over the frequency period to derive an annual cost share which will be the counterweight or offset used to measure the annual cost of security measures employed to reduce the flood impact.

By quantifying all the quantifiable security loss exposures in the manner noted and by calculating or estimating via the annual budget the cost of the security program, it is possible to produce a combined, estimated cost of security loss plus cost of security program. With these estimates a preliminary assessment of the security program's value can be made as follows:

$$\frac{ELBP - ELAP}{CSP} = ROI \text{ (return)},$$

where ELBP = Estimated losses before security program. To include all events for which probability/criticality have been calculated.

ELAP = Estimated losses after security program implemented. This will be those losses which will incur despite the program.

CSP = Cost of the security program. To include expense cost per annum and allocated share of capital cost.

This calculation is really only a forecast but it is a basis, even before investment has been made, to focus attention on the planned benefit. A security program plan which showed less ROI or less return on the investment than could be obtained by some other use of the resources would be subject to serious scrutiny on the grounds that the planning was poor or the program uneccessary.

At the end of the time period a more precise measure can be made because actual data will be at hand. In this situation the calculation would be.

$$\frac{(ELPB - ILAP) + AR}{CSP} = ROI,$$

where ELPB = estimated losses before program

ILAP = incurred losses after program

AR = actual recoveries through program

CSP = cost of program

In this case the difference between estimated probable losses before the program and incurred losses with the program, plus the amounts of recoveries achieved through the program such as diverted assets recaptured, successful proofs of loss developed against insurance coverage, claims or third party liability actions made possible, etc, will represent the total return of the program. Dividing that by the cost of the program over the same period will give the return on that amount of investment. The higher the ROI or ratio, the more valuable the program.

Diminishing Return

A moment's reflection will suggest that using the foregoing techniques, a truly effective program might show a decreasing return on investment in that actual or incurred losses will keep shrinking. However, it is always important to keep in mind that the vulnerability to losses is what the program effects. Given no program, the losses now avoided would have the original probability of occurrence. In a way this leads to a "zero base" approach to the security function. Each year the entire exposure should be recalculated on the basis of the then exposures and their probability/criticality relationships. As the incurred losses are reduced and remain low, the amount of avoided loss will still account for the real size of the actual exposures. The investment return will still be a valid measure.

REFERENCES

1. Webster's International Dictionary, 3d Ed.

2. Black's Law Dictionary, 5th Ed.

3. President's Commission on Law Enforcement and the Administration of Justice, *The Challenge of Crime in a Free Society,* (Washington, D.C.), Government Printing Office, 1967), p. 21.

4. Ibid. pp. 23, 24.

5. U.S. Dept. of Commerce, Bureau of Domestic Commerce, *Cost of Crimes Against Business,* (Washington, D.C.), Government Printing Office, 1976), p. v.

6. American Management Association, *Crimes Against Business: Recommendations for Demonstrations, Research and Related Programs Designed to Reduce and Control Non-Violent Crime Against Business,* New York, 1977.

7. The Small Business Administration, *Crime Against Small Business,* Senate Document 91-14, (Washington, D.C.) Government Printing Office, 1969), p.1.

Self Test Questions

1. What is the difference between a security loss and a competitive or speculative loss?

2. What is the function of grids and matrixes in security vulnerability assessment?

3. What are the classes of factors effecting loss occurrence probability or frequency?

4. Define "Loss Criticality".

5. Identify the usual categories of costs connected with security losses.

6. Explain the significance of a combined Probability-Criticality rating assigned to a loss event.

7. A useful formula for determining the economic benefit derived from a security program can be stated, $\dfrac{ELBP - ELAP}{CSP} = R.O.I.$

Explain the terms and the equation.

8. Name some positive types of recoveries achieved through security programs.

2

Organizing a
Security Operation

SOME of the more essential elements that need to be considered in the design and implementation of a protection program are discussed in this chapter. The discussion has been divided into six categories: (1) delegation of authority and responsibility, (2) organization, (3) staffing and training, (4) planning, (5) relationships, and (6) control. The vulnerability analysis discussed in Chapter 1 should be completed before any of the six items are considered, because hazards and risks must be analyzed and cataloged before action to cope with them can be taken.

A security program may not be readily acceptable to supervisors or employees because it limits the freedom of people and exerts some control over them. Also, it is an expensive overhead item that does not contribute directly to the production of the organization. A protection organization is difficult to administer because of those factors.

Top officers of an enterprise must be sincerely interested in having a security program; their complete support is essential if the program is to be successful. They cannot agree in a casual, tacit way that a security program is merely a good idea. Instead, they must personally become involved and show all levels of supervision as well as employees that the program will be completely supported and backed at the top level in the organization. Without that, the odds are great that any security program that may be designed will be doomed to failure. Next, all subordinate supervisors must become a part of the program because the program will not succeed without acceptance at all levels of management. The key to acceptance is understanding and knowledge.

Delegation of Authority and Responsibility

A basic first ingredient in considering a protection program is the delegation of authority from top management to the executive who will head the protection organization. The responsibilities the protection unit will assume should also be defined, usually by a policy statement or series of statements that may define the attitude of the organization toward the protection of assets. A statement that a security organization is being established to implement the program might also be included.

Further, the policy may also define the areas for which the unit will be responsible. That definition will result from the vulnerability analysis already mentioned. Some examples of questions that must be asked are: Will the protection organization assume responsibility for safety and fire protection? Will it be responsible for disaster planning and control? Some companies have combined all protection functions to include the regular security activities as well as related activities such as auditing and insurance administration. The resulting functional activity has been designated risk management or other appropriate term to reflect that the entire spectrum of protection has been integrated into one security unit.

Whether accomplished by a policy statement or otherwise, the areas of responsibility should be clearly defined and understood by the top management representatives as well as the individual or individuals involved with the protection program.

Organization

As is true of any operating activity, organization is not only basic to the planning of an effective protection program but essential if the program is to be well managed (Figure 3). As a result, after the tasks that the protection organization will perform have been defined, an organizational structure must be developed to discharge the delegated responsibilities.

A typical present-day security organization designed to provide complete protection for the average industrial facility performs a variety of functions. Examples of some functional areas are guard controls; internal theft control for the protection of documents, records, and negotiable instruments; controls for the prevention of espionage; fire prevention; disaster control; badge and identification; and security education of employees. All are essential for complete protection. "Security in depth" is a term commonly used to indicate a complete series of controls resulting from an effective organization.

The executive who has been delegated the authority to plan the protection program must be capable of assuming complete responsibility for developing the security organization. A basic concept that should be kept uppermost in mind is that the organization developed should be structured to meet the particular needs of the installation or enterprise it will be required to service. Therefore, the organization should not be copied from one used elsewhere unless it has first been carefully analyzed to insure that it meets the requirements of the installation where it will be utilized.

When the development of a security organization is planned, there are several factors that need to be considered: reporting levels, line and staff relationships, and the structure of the organization to provide for the protection of vulnerable areas as well as to provide for delegation of responsibility and authority within the protection organization. The remainder of this section on organization will be divided into those considerations.

Reporting Levels

The reporting level for the protection executive in the management structure is important if that executive is to be effective in discharging the responsibilities delegated. Generally speaking, individuals responsible for security programs should be at organizational levels such that they can operate as key figures in the administrative management structure of their enterprises. They should participate in the overall administrative planning and at all times be intimately familiar with the goals and objectives of their organizations as well as long and short-range planning. If the organization level is the same as that of the personnel executive, controller, and other like administrative functional representatives in the organization, the prctection executives will usually be in a position to participate effectively as a member of the key management staff.

If the protection executive is in a position to participate fully in top-level planning, a geat contribution can be made to the profit-and-loss statement of an organization. However, the individual must operate at a level in the organization such that a worthwhile contribution can be made. If, on the other hand, the executive is buried in the organization and is never privy to top-level deliberations, such an executive will not be able to function effectively regardless of the authority and responsibility that has been delegated.

Surveys that have been conducted reflect that increased importance is being given to the position of the security executive. In organizations that have well managed programs, the security executive usually reports to a senior executive not more than three steps down from the chief operating executive[1].

Line and Staff

Both staff and line organizations are utilized in the assets protection field. As a result, a short discussion of that area may be of value. "Line," as utilized in this chapter, refers to the protection unit at the operating level, at the plant or division level, for example. The line protection executive, then, at the operating level would ordinarily be responsible to the top executive of an operating activity and would supervise the day-to-day operation of the security function. The line executive would ordinarily have under his supervision an operating staff such as guards, investigators, and firemen.

Figure 3. One of the Earliest Accounts of Organization Planning—and of the Results Achieved Thereby—Is Found in the Book of Exodus

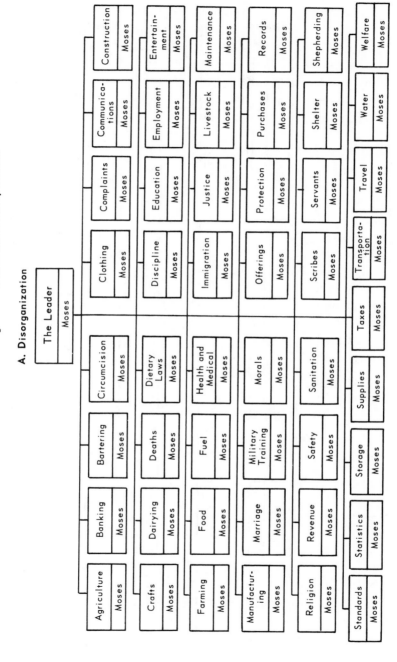

A. Disorganization

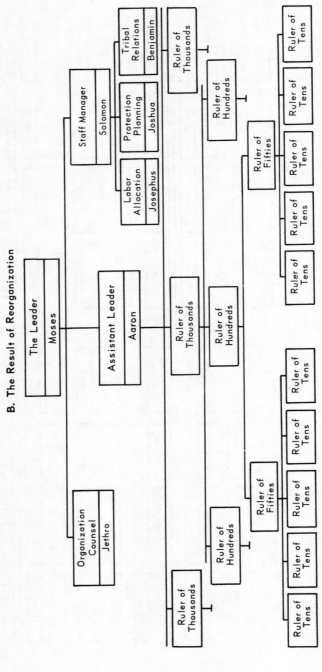

FIGURE 3 (Continued)

B. The Result of Reorganization

When the Israelites left Egypt, they had the simplest form of organization possible. There was one leader, Moses, and a mass of followers, as shown in (a). But that organization proved unwieldy. Moses, the Bible tells us, sat by the people from morning to evening, deciding disputes between them. In consequence, he and his followers made little progress toward the Promised Land in more than a generation. When a better form of organization was introduced, as shown in (b), the Israelites made as much progress in less than a year as they had in the preceding 39 years. SOURCE: Ernest Dale, *Organization* (AMA, 1967)

The terminology "line" and "staff" have a variety of meanings, as indicated in this quotation from a textbook on organization:[2]

> While business has adopted the words staff and line from military terminology, it has shown insufficient insight into the realities those terms are used to describe. In consequence, there has been almost endless confusion and conflict about the part that so-called staff positions are supposed to play in business organizations. Definitions of "staff" given by authorities on business management often tend to be vague, contradictory, and lacking in precision. The setting up of staff positions in practice has too frequently been accompanied by disputes about competence and authority which have often led to the discrediting, if not the dismissal, of able and enthusiastic younger executives. Communication in business—and it is to the system of authentic communication that the word "line" really refers—is still far less precise and orderly than it is in the combat forces. In short, business has taken over from the military nothing but the words. It is still making its own somewhat elementary and painful experiments with the problems of human relations involved in line and staff relations.

According to another textbook:[3]

> The staff executives are those who facilitate the work of the line, performing services for it, providing it with advice and information, and auditing its performance in various respects. In general, three types of staff are widely used: (1) personal staff, (2) specialized staff, and (3) general staff.

As there are many interpretations of the term "staff," the framework within which it will be used in this discussion should be noted. It is the specialized staff that will be considered. According to another textbook on organization:[4]

> Specialized staffs represent an authority of ideas instead of line authority to issue orders. Their job is sometimes said to advise other units, but this is only one of three types of activities they have, as follows: (1) advisory—the staff guides others; (2) service—the staff performs work for others; and (3) the staff regulates and constrains others.

An example of the staff executive as utilized in the assets protection field is the director of security, found in many major corporations, with staff responsibility for plants, divisions and other facilities located throughout the country and perhaps in foreign countries. The security executive at each of the operating activities within the corporation ordinarily would perform as a line executive in such a situation.

The corporate staff director normally assumes responsibility for the protection program for the entire enterprise, and then by means of policy statements defines the requirements of the corporate program for all the line executives. Policy has been defined as what to do, and not how to do it. Upon receipt of the corporation's direction, the line security representative at the operating level would, then, implement policies defined by the corporate staff executive through procedures, practices, or other how-to media.

The security staff director at the corporate level will ordinarily report to a top management representative in the corporate headquarters—often a vice-president—and the line executive will usually report to a key management representative at the operating level. The head of the operating activity exercises general supervision over the daily performance of the line executive responsible for the protection program. The corporate staff director has what has commonly come to be known as a dotted-line relationship with the security executive in the operating activity. The executive at this level does not exercise direct supervision over the line executive, but does have staff responsibility, as a representative of the chief executive officer of the corporation, to insure that the line security organization is operating within the framework of corporation policy and that the protection program at the operating level is effective.

If the corporate director should determine that corporate requrements were not being met and that the line activity was deficient in some way, the corporate executive would normally inform the line security executive and make suggestions for correction. That should solve the problem. However, if the deficiency continued to exist, the staff director could next bring the problem to the attention of the supervisor of the line security executive. If the negotiations with members of supervision at the operating level should not produce the desired results, the corporate staff executive as a final action could obtain the assistance of the executive at the corporate level with line responsibility for the operating activity and have that executive, either verbally or in writing, direct that the necessary corrections be made at the operating level.

In the case of the short discussion of the situation just outlined, the corporate protection staff executive would be performing all three types of activities mentioned for which the specialized staff representative is responsible: advisory, service, and control.

Corporate directors usually has a relatively small group of staff specialists working under their supervision, and line executives ordinarily have a large number of subordinates, such as guards, firemen, and investigators, under their direction to implement their protection programs. Usually, the larger the company, the larger the staff group will be. In a smaller company there may be no staff function. However, that does not mean that staff activity such as policy direction is not required. Instead, it will be performed as part of the line or operating job, so that the executive in the smaller company may perform the staff as well as the line function.

Organization Structure

The following is a basic discussion of organization design and theory:[5]

> Organizational engineering is an ancient calling. Aristotle, Machiavelli, Bentham, and Madison devoted considerable energies to the design of purposive organizations. In more recent times, Urwick, Taylor, Fayol, Mooney, Gulick, and Davis focused on the problems associated with engineering joint human effort through complex organizational systems. Out of this work has come a relatively distinctive vocabulary and some principles of design. The dimensions of organizational anatomy have been reduced to a few simple canonical symbols. In the purest form of descriptive anatomy we require only two symbols: a rectangle (representing a basic element of the organization—a position) and a directed line (representing a relationship between two positions). The relationship is normally defined in terms of such expres-

sions as "reports to," "has authority over," "is the boss of." By using these two symbols, we can construct a schematic representation of an organization—the familiar organization chart. Alternatively, we can translate the basic notation into one appropriate for graph theory or matrix manipulation. In any case, much of the basic material of classical organizational engineering represents constraints imposed on this simple chart.

In designing the organizational structure, the functions that need to be performed to cope with the vulnerabilities should be carefully examined to insure that all the hazards are taken care of. Related functions should be grouped or combined into subunits so that the span of control is limited and the executive in charge of the overall organization can effectively direct and supervise the protection program through his subordinate supervisors. V. A. Graicunas, a Lithuanian management consultant who is often quoted on the subject of span of control, explained in 1933 that one person should not supervise the activities of more than five or at most six other individuals with interrelated responsibilities. The basic theory on the short span of control is probably taken from a British general, Sir Ian Hamilton, who wrote in 1921:[6] "The average human brain finds its effective scope in handling three to six other brains."

The following caution about adopting such a formula arbitrarily should be noted:[7]

Formulas of the type just described are useful in showing proportions, but they should not be taken as a literal representation of each manager's relationships. Many other factors determine the number of employees that one person can effectively manage. Some of these are capacity and skill of the manager himself, capacity and skill of employees managed, stability of operations, contacts with other chains of command, type of work managed, and time devoted to community activities. The result is that spans range widely from one to one hundred or more direct subordinates, with somewhere between five and fifteen being typical.

Delegation

After the organization structure has been determined, it is essential that responsibility and authority be delegated to each subunit. Delegation has been defined as the assignment of duties, responsibility, and authority to someone else.⁵ If the assignment is accepted, the individual then becomes the representative of the executive who gave the assignment. If the assignment of responsibility and authority is not accepted, the delegation is not successful. Executives can extend their influence beyond their personal time, energy, and knowledge only through successful delegation.

Some security executives are incapable of delegation. That the problem is not peculiar to the protection area is indicated by one study of the general business management area that indicated that inability to delegate caused the failure of more managers than any other shortcoming. Some executives cannot psychologically allow themselves to delegate because they feel they may be giving something away and that it will weaken their positions. Others have no confidence in depending on others to perform work for which they are held ultimately responsible. However, the security administrator must realize that delegation initiates management and that if there is no delegation, no one is managed.

Delegation is particularly important if the protection organization is to be effective because it must operate 24 hours a day, 7 days a week. Serious problems can, of course, be expected at any time, day or night. If the top executive and the subordinate supervisors do not delegate, a bottleneck will almost surely result because they cannot possibly be available at all times to make decisions. An accepted management theory is that tasks are delegated in accordance with the job to be accomplished. Executives and workers are hired and trained to perform jobs; tasks are usually not changed to fit the people available.

After delegation of responsibility and authority within the protection organization has been clearly defined, the information can be reduced to writing and disseminated to members of the organization. The material can take the form of internal instructions or procedures and should be in sufficient detail that those working in each functional unit clearly understand their responsibilities and their relationships to those in other functional units.

Staffing and Training

In years past when the need for assets protection began to be apparent, the classic solution to the problem of staffing the security organization was to select individuals whose background and experience were limited to law enforcement. Many organizations regarded security as a low-grade responsibility. That is understandable, because the administration of security as it is known today originated as an activity limited to plant protection with the emphasis on the watchman or guard.

An evolution has taken place over the years until the planning and direction of the security program in many organizations is now regarded as an essential and specialized management area. It is no longer limited to plant protection but must be a complete activity with a great variety of functions if the assets of the modern facility are to be given adequate protection. Although the guard activity is usually an important function in any security organization, it is ordinarily but one element in the modern protection program. Some law enforcement experience may be needed because of the plant protection activity, but it is no longer a prime requirement for those being considered for supervisory positions in the present-day security organizations.

Investigators from government investigative agencies have also been hired as security administrators. That solution has not always resulted in success because, as is true of individuals with backgrounds limited to law enforcement, those with successful records as investigators are not always effective as administrators.

The comments about law enforcement or investigative experience are not made to discourage employment of administrators with that type of background, because there are many with one of those qualifications who are doing outstanding jobs as protection executives. However, it should be emphasized that experience of that type alone should not be regarded as qualifying an individual for a security administrator's position.

Some special skill is needed to handle the administration of a security program, but no more specialized knowledge than is required to manage other administrative activities. To be successful, the contemporary security executive must be able to utilize the methods and techniques that comprise up-to-date management sciences. A codification of standard job qualifications and titles has never been developed. For that reason, successful executives in the security field are former engineers, attorneys, educators, military officers, or general business managers or represent a myriad of other backgrounds. To repeat, some security knowledge or experience may be helpful, but a more important qualification of the security executive will be ability to utilize modern basic management skills.

As the field of assets protection has gradually developed over the years into a more demanding administrative function, the prerequisites with regard to education and training have become more aligned with those of other administrative management positions. As a result, persons entering the security field in recent years have generally had college degrees. Those who have accepted top security administration positions have usually had that basic education qualification plus several years of specialized experience and training in administrative management.

Compensation

Compensation for security administrators will vary as for any other position. Because of the earlier orientation toward plant protection and the emphasis on watchmen or guards, the pay range was less in earlier years for those administering programs of that type than is required for the executives directing more complex present-day programs. Because of the earlier plant protection influence, compensation for security executives has not kept pace with pay increases in other administrative fields. As the need for an effective security program has come to be recognized in many industrial organizations, it has followed that top managements of those organizations have recognized that it is necessary to provide adequate compensation to obtain and retain qualified administrators.

However, the salary level of the security administrator still lags behind in some business and industrial areas, so top management of an enterprise requiring top-level administrators should not be misled by a general wage survey of other organizations. The low salary ranges in some companies paid to security administrators who are not capable of utilizing modern management techniques may give a false impression of the pay range that should be established. A better way to approach the problem of establishing compensation for the security administrators is to base any wage survey only on salaries paid to executives in companies with highly effective programs. If that is done, it will usually be found that the salary ranges established will be much higher than the general average would show in a wage survey. However, that should insure a compensation plan to attract above-average administrators who can cope effectively with the complexities of the present-day protection program.

As this was being written in 1981, it was estimated that the compensation varied from about $30,000 per year for an executive administering a small corporation or division program to more than $90,000 per year for executives directing giant corporation programs. More and more administrators at the top level are being appointed vice-presidents. At the next level, the administrator of a small division or plant or a first-level supervisor under a director of a larger corporation or division, the compensation varied from about $25,000 to about $40,000 per year. At the next supervisory level, the compensation varied from about $20,000 to about $30,000 per year. The actual salary paid each individual varies, of course, with the size of the organization and the complexities of the position.

Education and Training

Education and training for administrators is now recognized as an essential element in any successful security program. For example, a 1979 survey conducted by the American Society for Industrial Security revealed that there were more than 180 colleges and universities offering a·minimum of one security course compared to only 43 institutions offering such courses in 1972. The survey further reflected that an increasing number of institutions are offering degree programs in the security field.[9]

Further, the steady increase in attendance at training programs sponsored by the American Society for Industrial Security, which has become the undisputed leader in the presentation of high quality training programs, indicates there is great interest in security management education and training.[10]

A typical high quality course offered by this society is the "Assets Protection Course", a four and one-half day meeting, which has been offered twice a year in the United States and periodically abroad. Currently regarded as the only top quality basic security course available for newcomers to the field as well as a review course for experienced professionals, the course has usually been sold out well in advance of the start date.

Another course designed for top level security executives, also sponsored by the American Society for Industrial Security, is the Advanced Security Management Program. The attendance has been limited to twenty-five participants who are required to qualify as being key management personnel in charge of major programs. The course is an intensive, prestige, five-day learning experience which offers attendees an opportunity to study and apply financial, political, social and psychological principles which are required for effective security management at the top level in the dynamic environment of the modern enterprise.

The American Society for Industrial Security also sponsors a variety of one or two day workshops each year as well as the Professional Certification Review Course. This is a short practical review of eight mandatory subject areas in a written proficiency examination required to be taken and passed to obtain the designation Certified Protection Professional, CPP.

Because of striking technical advances in all areas, the continuing education of security executives cannot be limited to security and management subjects. The expanding use of electronic data processing is only one example.

In order to keep pace with such technical advances, security executives must become aware of how modern up-to-date techniques and equipment can be efficiently used in the protection program. They must also become aware of uses in other areas in the enterprise so that the security program can be planned to take into account the new techniques and technical advances being utilized throughout the entire organization.

Planning

Planning is also an important element in the administration of the protection program. Since security is essentially preventive in nature, it is necessary that future risks and hazards be anticipated as much in advance as possible. If they are not, the security organization will constantly be forced to improvise in reacting to problems. Some protection programs are administered in precisely that way, a type of administration that has been described as crisis management.

General Planning Objectives

As a first step in the planning process, the objectives and the overall planning for the future of the entire enterprise must be understood and related to the protection program. General objectives for the security organization should also be established. Those objectives should be considered when the planning is being done. The planning of the protection program must be realistic, and it must be kept in mind that the basis for the planning of the complete enterprise is profit and loss. Therefore, close attention must be given to that factor in protection planning. All plans must be related directly to costs.

What period of time should planning encompass? The protection program should be planned as far in advance as possible. Some projects may take several years to accomplish. For example, an automated document control system utilizing electronic data processing and computers might take several years to completely implement, and an automated personnel control system utilizing the systems approach might also require several years to become fully operational. As another example, a complete emergency plan to cope with disasters will normally take a considerable period of time to develop and implement. Therefore, a 5-year planning cycle might be considered as the basis for the planning effort. The plan for each project to be undertaken during the 5-year period can be broken into segments and laid out by year until completed.

Project Planning by Year

The next phase in the planning program might then be a detailed plan for the next year of operations. Enough is generally known of requirements a year in advance that detailed projects can be defined to be completed during that current year. In the more detailed timetable, it is best to not only define the overall project but also assign a completion date for it. In addition, a single goal or project can be broken into phases or segments and anticipated completion dates can be attached to each of the parts. Everybody has a tendency to procrastinate. By planning, those who administer the protection program can force themselves to look ahead to accomplish goals. Also, the plans made should be reviewed periodically to insure that progress is being made. Otherwise, the realization that a goal has not been met may come late in the year and after the opportunity for meeting it has passed.

Finally, the plans made should be translated into money and personnel needed. If the plans have been carefully thought out and well done and if they directly relate to the needs of the enterprise, the need for any budget increases will be amply supported by the plans and the entire package will be a logical presentation for top management approval.

Appraisal of Results

Quantification of results is another benefit that can develop from the planning process. The executive in charge of the security program and subordinates will know the current status of each project. As a result, the progress of the entire protection organization can be appraised. At the same time, supervisors of subordinate units within can be measured. Also, an appraisal can be made of how well the delegated responsibility previously discussed in this section is being accepted by subordinates. If a subordinate is not obtaining results, it may become apparent he is not accepting responsibility delegated to him.

"Management by objective" is the designation often given this type of arrangement. The top executive first delegates authority and responsibility for the accomplishment of projects and then through the planning process checks periodically for results. Accountability, therefore, must involve the individuals responsible for the supervision of each element of the protection organization. Compensation should also be tied to performance so that individuals are aware that if they consistently do not meet objectives, their pay checks will be affected.

Of course, deadlines must be flexible. As a part of the periodic planning review process, the protection executive must consider mitigating circumstances. If the program is reviewed frequently, difficulties should become apparent early enough that schedules can be readjusted or other changes made to compensate for unforeseen problems that can always be expected.

As a result, the top security executive as well as subordinates will be able to constantly look for deviations and should be ready to stimulate other actions or make changes as necessary. The protections executives at all levels will know what is going on in the organization at all times because they will constantly be reviewing results. Since the regular problems will constantly be worked on, on a routine basis, the executives should be ready to concentrate on large unanticipated problems when they occur.

Relationships

There are two general categories of relationships that are important in the successful management of the protection program. The first is internal and has to do with top-level management and employees in the organization. The second is external and it involves relationships with other security organizations; government agencies, including federal, state, county, and municipal; and activities in societies, associations, and so on.

Internal Relationships

As was pointed out under "Reporting Levels" earlier in this chapter, executives responsible for protection programs must be at an organizational level such that they can participate in top management planning and decision making. The protection executive must then become an integral part of the management team and take full charge of all protection responsibilities. In this connection, they must make the top management aware of hazards faced by the organization so that management support can be obtained in taking preventive measures. Therefore, the protection executive must be effective in indoctrinating top management so that those representatives will not only be aware of programs being undertaken but will also be aware of the value of a well-administered protection program.

Relationships with all other departments such as finance, plant engineering, and marketing are important in implementing protection programs. Arbitrary decisions or decisions that are unilateral can be disastrous regardless of their intended ultimate value to the organization.

In considering relationships it is well to remember that security is difficult to administer because by nature it limits the freedom of people or in some fashion controls them. As pointed out earlier in this chapter, security is not readily acceptable to management or employees, and it is an expensive overhead item that does not contribute directly to the production of the organization.

For those reasons, the inputs of other departments within the organization are important. For example, both the public relations and industrial relations viewpoints should be obtained for any program that might have an effect on employees. If an expenditure of money is being proposed, the impact should be discussed with the appropriate representatives in the finance unit. Other organizations in the enterprise should become involved, and their support and approval should be obtained before an idea or concept is presented to top management. It might, for example, be a useless exercise to present an idea to top management for approval if the controller would give a negative report when asked about funds available for the project. Therefore, new ideas or changes must be thoroughly researched and considered. Then the idea or plan must be discussed, perhaps informally, with other affected departments or groups within the company to insure their understanding, cooperation, and support.

After that has been done, the idea or program is ready for presentation to the top officers of the company. By that time the presentation should include the background, an outline of the reasons for the new program, cost and savings analysis, criticisms and suggestions from affected departments, and a clear-cut action program by which the idea can be implemented.

Many good ideas and programs are never implemented because they are never clearly understood by the top management representatives who must approve them. A presentation of an idea or program to top management should cover every aspect of the problem and the solution; in other words, it should represent completed staff work. The presentation must be keyed to the company situation at the time of the presentation. If cost savings are uppermost in management's mind, savings should be emphasized in the presentation. If company expansion is a current planning concern, the presentation must emphasize how the proposal fits into and facilitates an expansion program.

To ignore matters of immediate concern to management, regardless of how irrelevant they may seem to be to the matter at hand, is to invite an immediate rejection of an idea. Indeed, if a program is such that it cannot be keyed to management's present concerns, consideration should be given to shelving it for a few months pending the arrival of a more auspicious occasion or a reworking of the plan to incorporate a definite consideration of management's immediate concerns.

Off-the-cuff presentations can defeat good ideas. For that reason careful consideration should be given to the actual presentation of the proposed program. Whenever possible, the presentation should be done with charts. There are a variety of ways of preparing charts, and they need not be elaborate. They might be hand-drawn with felt pens, or they might take the form of Vu-graphs. Regardless of how they are made, visual charts present facts and figures in a simple, easily grasped way that will make the presentation more readily understood and accepted.

With the best preparations, the result hoped for may not always be forthcoming; but even if the program does go down to temporary defeat, two gains will result. In the first place, the busy top officers will be aware that a complete job of staff work has been done and the problem has been presented objectively and in a businesslike way, together with a workable solution. They will appreciate that the proposed program has already been discussed with affected departments and that as much as possible has been done to meet objections and gain cooperation. The second gain, of course, is in learning what objections management has to the program. Is it cost alone, or is it cost plus other factors, such as anticipated morale effect? What is learned through the rejection of the program will enable the executive to see the program through the eyes of the objectors and to make such changes or collect such facts as may be needed to meet the objections in preparation for another presentation at a later time.

A great deal has been said and written about the problem of selling security to management. As already pointed out in this section, security is difficult to administer because it limits the freedom of people and is an expensive overhead item. A periodic sales pitch to top management will not gain acceptance if the security executives are not a part of the management team and are not utilizing the management techniques described in this chapter. A successful security program cannot be implemented simply by selling it to management. Performance of the executives and the organization is what counts with top management in the final analysis.

External Relationships

Good working relationships with organizations external to the enterprise are as important as the relationships within. That includes other business organizations as well as government agencies at all levels from the federal government down to the municipal organizations in localities where the organization does business. Because protection problems may involve any of the outside organizations mentioned, it is important to develop appropriate relationships in advance of the development of problems.

To wait until there is a serious security problem to build a good personal relationship with an external organization is to wait until it is too late. In addition, it is important to know in advance the identity of the individual in such an organization who can most effectively handle a particular problem. A great deal of time and effort can be saved if the proper relationships are developed as a part of the day-to-day management of the security organization.

How can such relationships be established? An excellent way is to be active in professional organizations, which usually have business meetings, luncheons, and so on. Attendance at such meetings will enable the security executives to establish good contacts and become personally acquainted with others who might give assistance or information in connection with a protection problem.

The top executive in the department cannot usually develop all of the contacts for the whole security organization. However, if all the supervisors in the protection organization are assigned the responsibility for establishing relationships in their various areas of interest, it will usually be found that sufficient external relationships can be established that assistance with almost any security problem can be assured.

Control

"Control," as used here, refers to techniques of management utilized to insure that established procedures are being followed not only in the security unit but through the entire organization. Also, it refers to techniques used within the protection organization to insure that security personnel are properly administering the program.

An inspection and audit system can provide the means of determining the effectiveness of the protection program throughout the entire organization. To implement such a system it is not necessary to have a group of auditors or inspectors who function only in that role. All administrators in the security organization can be assigned such tasks as a part of their regular duties. Also, those in the management chain in the organization of the enterprise can assist if a system is so designed they can participate. For example, a system of periodic self-inspection can be an effective tool with which supervisors can assist in the protection program by checking the security performance of their units and their employees.

Periodic review meetings can be scheduled with various levels of supervision, at which time the security performance of the organization can be discussed. That technique will not only interest supervisors in the protection program but make them a part of it. Consequently, they can assist the security executives with the administration of the program.

The security program must be constantly reviewed by the security administrators, and changes and revisions must be made as required. Otherwise, control can be lost. Changes in company facilities and product lines or other revisions may create areas of vulnerability that require changes in the protection program.

Within the security unit, conferences can be scheduled with individual supervisors on a routine basis to review the progress of each protection subunit and discuss problems. Also, group meetings of all supervisors in the security organization can be scheduled to hear individual supervisors report on their activities in the unit. If planning is formalized, as suggested earlier in this chapter, the projects and their deadlines can serve as a basis for the discussions with individual supervisors as well as the group meetings of supervisors.

Reports, both statistical and narrative, can be required of executives in the security organization. However, there is a danger that such reports will be meaningless recitals of routine facts and statistics. If the reports are to be meaningful, they must interpret facts and statistics and, in effect, give meaningful information on the status of the protection of assets in the enterprise.

REFERENCES

1. Dun's Review, *Corporate Security: Top Management Mandate,* January 1980, page 94.

The Wall Street Journal, *Security: The Essential Corporate Asset,* August 10, 1978. Special supplement on security.

Fortune. *Security: A Concept and a Management Technique,* September 1974, page 47.

2. Ernest Dale and Lyndall F. Urwick, *Staff in Organization* (New York McGraw-Hill book company, 1960). Used by permission.

3. Ernest Dale, *Organization* (New York: American Management Association, 1967)

4. Keith Davis, *Human Relations at Work: The Dynamics of Organizational Behavior* (New York: McGraw-Hill Book Company, 1967). Used by permission.

5. W. W. Cooper, H. J. Leavitt, and M. W. Shelly II, *New Perspectives in Organization Research* (New York: John Wiley & Sons, Inc., 1964). Used by permission.

6. Ernest Dale, *The Great Organizers* (New York: McGraw-Hill Book Company, paperback, 1971). Used by permission.

7. Davis, op. cit.

8. Ibid.

9. As reported in *Security Management,* published by the Americn Society for Industrial Security, January 1980, page 43.

10. For additional information about training sponsored by the American Society for Industrial Security, contact the Society Headquarters, 2000 K Street, N.W., Washington D.C. 20006.

Self Test Questions

1. What are some of the items that should be included in a delegation of security policy statements issued by top management?

2. At what management level in the organization should the executive in charge of the protection program report?

3. What should be the expected result in a 24-hour a day, 7-day a week security operation if delegation is not practiced?

4. Although the classic solution to the problem of staffing a protection organization in the past was to obtain individuals with experience limited to law enforcement, what qualifications and experience are now more appropriate in a modern progressive enterprise?

5. What are two general categories of relationships that are important if a protection program is to be successful and why are they important?

6. Why is it important, when developing any security organization, that an organization utilized by another enterprise not be copied without giving consideration to the requirements of the enterprise requiring a protection organization?

7. List some "control" techniques that can be utilized to insure that established procedures are being followed.

=== 3 ===

Essentials of a
Security Program

THREE essential elements should be included in the design of an effective protection program: education, prevention, and detection. The first two are the most important because risk avoidance is the key to good security protection. Therefore, through the design of a good education program and the implementation of good preventive measures, many of the risks that otherwise might exist can be avoided. An effective risk avoidance plan offers these benefits:

- Individuals who would be tempted to perform acts that would damage the enterprise will be discouraged because the odds that they will be discovered are great.
- Problems that might cause serious losses will be discovered in the early stages.
- The cost of insurance will be kept at a minimum because of a favorable experience rating.
- Information used in making decisions will be more reliable. For example, falsification of vital information to conceal

thefts or fraud will be less likely, and as a result top management will be in less danger of basing important business decisions on erroneous information.

Detection is important for two reasons. First, management of an enterprise should be aware as soon as possible that an event that might cause damage has occurred so that appropriate controls can be instituted to prevent a recurrence. Second, those who would be motivated to perform acts that would be detrimental to the organization will be aware they may be caught.

Although detection is important, it must be remembered that severe damage may have resulted by the time the problem is discovered. In fact, damage might be so severe that the future operation of the organization might be seriously affected. The enterprise, of course, has legal protection and can take action against the individual or individuals who are responsible for damage under both criminal and civil laws; but although the organization may obtain an award for damages under civil law, it often happens that it is impossible to collect a significant part of the awarded amount. Often the only result is that the individual or individuals involved receive fines and/or sentences under the appropriate criminal statute. That, of course, is no compensation for monetary losses the organization may have experienced.

Education

An education program should be planned to include three groups in the organization: the top officers, supervisors at all levels, and employees. As was pointed out in an earlier chapter, a protection program must be completely accepted by the upper levels of management. The top management representatives should understand completely the risks the enterprise will face if it has no security program and be familiar with the steps implemented to protect the assets of the organization. In addition, by their actions and the interest they show in the program, top management personnel must indicate to all levels of supervision and employees that the program is supported fully at the top level.

All subordinate levels of supervision must become a part of the program because a protection program cannot succeed without acceptance at all levels of management. The key to acceptance is understanding and knowledge of the program, and that is true of employees as well as management personnel. Ordinarily, if an individual understands the reasons why a security requirement is being implemented, he will accept it. If employees do not believe in the protection program and do not want to cooperate, the program cannot be enforced by security personnel alone regardless of their number. The effect of a lack of voluntary acceptance of a protection program is much like that of depending only on the police for law enforcement in cities:[1]

> The public peace—the sidewalk and street peace—of cities is not kept primarily by the police, necessary as police are. It is kept primarily by an intricate, almost unconscious, network of voluntary controls and standards among the people themselves, and enforced by the people themselves. In some city areas . . . the keeping of public sidewalk law and order is left almost entirely to the police and special guards. Such places are jungles. No amount of police can enforce civilization where the normal, casual enforcement of it has broken down.

Employees must be motivated to want a security program. Awareness of the need for a program and an acceptance of it are essential. For that reason a security education program is the cornerstone of an effective assets protection program. The program must overcome the natural reluctance on the part of all employees to accept a protection program. Employees must become a part of the program and not grudgingly accept it as a necessary evil while giving it lip service, or, worse, fighting it.

A great deal has been written about the subject of security indoctrination and education of employees, and there are many factors involved. The problem is complex simply because it involves a complex subject—human beings. Here it is possible to discuss only some of the more important elements that should be considered in the planning of any protection program.

The education of employees may be formal or informal depending on the size of the organization. Regardless of how it is conducted, the program should be designed to give each employee a full explanation of the operation of the protection system together with the reasons a security program is required. In addition, employees should be told the specific responsibilities that each must assume for the protection of the company's assets.

An indoctrination program must not be limited to a lecture on security for each new employee. Instead, an effective program should include, among other things, the use of the company newspaper, bulletins, and periodic discussions in meetings. Also, the overall security policies as well as the procedures governing the responsibility each individual is required to assume for the protection of assets should be defined in writing. If the organization has adopted a policy prohibiting employees from accepting bribes and kickbacks, that policy should be published and distributed to all levels of supervision as well as to employees. Also, employees should be clearly informed in writing that theft, regardless of the amount, will not be condoned. The effect that dishonest actions can have on the company and how that relates to profits and, in turn, the employee's position as well as his security in the organization should be highlighted. Employees should also be informed of the corrective or disciplinary action they can expect if they violate security policies and practices.

To be effective, a security education program must constantly emphasize the responsibility of each employee for the protection of the assests of the firm. As a result of the program, individuals should come to regard security as an essential element in their job performance.

The security staff can assist top management and supervisors by developing a system to provide for the correction or disciplining of employees who violate security procedures. In that connection, the procedures adopted should provide for all corrective actions to be taken within the supervisory chain. Security personnel cannot be effective in attempting to correct an employee who has violated a security procedure. Such a problem should be referred to the supervisor responsible for the work performance of the employee. Security problems should be handled in the same way that other employee performance problems are handled.

The security executive may act as an adviser to the responsible supervisor in such cases and assist in the actions taken to correct the problem. If the supervisor to whom the problem is referred is reluctant or refuses to take action as required, the system should provide for the referral of the matter to a higher-level supervisor. In some instances, problems involving employees may be so serious that the top executive officer of the organization should personally become immediately involved.

A plan for the handling of employee problems can be so designed that there is a definition of how supervisors are to become involved. At the same time, a review procedure can be developed within the management chain to insure that problem cases are reviewed by the top-level executives in the organization. In that way, top management will not only become aware of protection problems in the organization but have an opportunity to react to corrective actions taken at the lower levels of supervision. However, it cannot be stressed too much that any protection program that does not utilize the management chain in this way and depends on security personnel to take corrective actions involving employees will probably fail.

Prevention

Applicant Screening

Although the value of good education and indoctrination techniques cannot be overemphasized, one preventive measure that should precede the education of employees is the screening and investigation of applicants. If a mistake has been made in the selection of an employee, the best education program possible may not prevent that employee from causing a serious loss.

The primary objective of a screening program should be to obtain and retain a reasonably suitable, trustworthy, and competent work force. It should be made up of individuals whose present personal lives and backgrounds conform to the accepted modes of the times in the community in which they work and live.

Responsible management representatives as well as reliable employees do not want to work in a facility with people who have repeatedly been involved in serious crimes or who have demonstrable violent tendencies. Convicted embezzlers, thieves, and shoplifters should not be employed in the cashier's cage or in the finance branch of the company. Alcoholics and narcotics addicts are a hazard to themselves and others, as are persons with uncontrolled or incapacitating nervous disorders. No political extremists of any kind have any place within the perimeters of a defense plant. Individuals of these types are often bad business risks as employees; they represent a poor investment.

A screening program is as important for the company involved in hiring the economically disadvantaged as it is for those not participating in that program. Since the economically disadvantaged may have known criminal records, the question might be asked why it should be necessary to investigate them. There are two reasons. First, it is important that the organization employing an ex-convict be aware of the extent of his criminal record. Second, an investigation should develop information that will assist the organization to assign the individual to a task or to an area where temptation will be at a minimum.

Because of the extreme importance of this aspect of risk avoidance, investigations are discussed more completely in a later chapter.

Physical Security Controls

Utilization of physical security controls is important as one element in the prevention of loss. Physical controls at all openings in a facility will be effective in preventing individuals who would cause harm from entering the facility. Physical barriers or controls within the facility are also effective in preventing those within from causing damage. "Security in depth" is a term commonly used to indicate a complete series of controls to insure adequate physical protection of a facility. Examples of physical controls are (1) guards, (2) alarm systems, (3) locks, (4) doors, turnstiles, gates, and so on, (5) barriers such as fences and walls, (6) lights, and (7) safes, vaults, and other special constructions.

Physical controls have been described as impediments designed to deter the undetermined and delay the determined. They also act as psychological deterrents by providing obstacles to penetration. Because of the controls, trespass should appear to be unprofitable in terms of gain or risk. However, physical controls are only one element in a complete security program.

Transfer of Personnel

Transferring personnel within the organization or rotating them periodically is an effective device for preventing a breach of security. If employees are aware that their job assignments may be shifted, they will be less likely to perform acts that might be discovered after they are transferred. Transfer is not, of course, always possible, but employees in sensitive positions might be rotated from one job to another. For example, branch managers might be shifted from time to time to different offices. The routes of truck drivers can be changed. Foremen and supervisors might be rotated within a factory. Clerks and supervisors who handle the financial affairs of the organization might be shifted within the organization periodically.

Emergency Planning

Emergency planning must take into consideration a variety of emergency situations. An emergency plan will not necessarily prevent the occurrence of an emergency situation such as one resulting from a natural phenomenon; but if actions are taken to cope with an emergency situation as it develops, an ordinary emergency may be prevented from developing into a tragedy. An emergency plan should provide for orderly actions and decisions so that loss is minimized.

Because problems involving civil disorders and terrorist activities as well as bombs are very important, individual chapters are devoted to those two subjects

Insurance

Although insurance cannot be depended on to mitigate losses completely, it is one factor in any loss prevention plan. Insurance protection is usually obtained by most organizations for coverage against property damage, business interruption, and liability. However, the insurance coverage should be carefully examined to insure that it will be effective in preventing large losses. For example, fire coverage should be reviewed to determine if it limits coverage by insurrection exclusions in the event of civil disturbance. It should also be determined if the insurance in force covers losses from looting that might result during a riot.

Insurance should not be relied on to cover all risks. Each risk should be examined to determine if the money that might be spent on insurance could more effectively be applied to security countermeasures designed to prevent loss. An adequate protection program can be achieved only through a balanced combination of security and insurance.

One type of insurance that is sometimes overlooked as a loss prevention measure is the fidelity bond or honesty insurance.[3] Because that type of coverage is important as a protective measure and because its value is often not understood by top management representatives, the remainder of this section is devoted to it.

Honesty insurance usually indemnifies the firm, to a stated limit for loss of money or other property occasioned by dishonest acts of its bonded employees. The bond covers all fraudulent or dishonest acts, including larceny, theft, embezzlement, forgery, misappropriation, and wrongful abstraction or wilful misapplication, committed by employees acting alone or in collusion.

Whether or not employees have access to the funds of a firm and whether or not they are authorized to buy, sell, ship, or store goods, they should be bonded in an amount calculated to offset potential thefts. The larger the firm's assets and the greater the turnover in volume of business, the more probable that a large loss or series of losses may be concealed over a long period and consequently the more urgent it is that the firm safeguard against such losses through fidelity bond protection.

A question that has frequently puzzled management of business and industrial firms is how much honesty insurance it is necessary to carry. The rising theft trend has revealed a serious condition of underinsurance, and therefore the answer to the question is of utmost importance. A careful study of employee dishonesty was made by The Surety Association of America. It revealed that in 65 percent of the actual losses studied over a 10-year period, the insured did not carry a fidelity bond large enough to cover his loss.

Among the factors governing exposure that were studied in the survey were total assets, goods on hand, annual gross sales or income, the nature of the business of the insured, the size of the firm, and the number of employees. Those factors were related to the amounts that had been stolen in a large number of actual instances, and they were utilized in developing a formula designed to gauge the minimum coverage appropriate for a commercial firm of any size.

The formula so developed employs an exposure index for determining the minimum amount of honesty insurance, or fidelity bond coverage, necessary for the insured. The index is not an estimated amount of actual exposure but a weighted base related to exposure to which the recommended bond amount is keyed. The exposure index is derived from two principal elements of exposure to large dishonesty losses: (1) current assets and (2) gross sales or income. By adapting the formula to the employee dishonesty exposure of a firm of any size, the minimum amount of fidelity bond coverage that the firm requires can be determined. The formula, which is applicable to any size or type of commercial firm, is given in Table 2 and the two charts in Figure 4 can be utilized to determine the amount of honesty insurance required.

In the final analysis, it is the employer himself, in consultation with his insurance agent, who can best determine his loss potentialities due to employee dishonesty factors and the coverages necessary to protect himself. He can be aided by his certified public accountant's understanding of his operations and internal controls and by the application of the formula of Table 2 to his business structure.

Table 2. Formula for Determining Amount of Honesty Insurance Necessary

1. Total current assets (cash, deposits, securities, receivables, and goods on hand).	$_____	
A. Value of goods on hand (raw materials, materials in process, finished merchandise or products).	$_____	
B. 5 percent of A.		$_____
C. Total current assets less goods on hand, that is, the difference between 1 and 1A.	$_____	
D. 20 percent of C.		$_____
2. Annual gross sales or income.	$_____	
A. 10 percent of 2.		$_____
Total of 1B, 1D, and 2A, the firm's dishonesty exposure index.		$_____
Suggested minimum amount of honesty insurance.		$_____

The grave peril of underinsurance is made obvious by the size and frequency of large-scale losses. The fidelity bond should be sufficient in amount to protect the firm against disastrous losses, and all employees should be covered, because employee dishonesty is not limited to those who hold trusted positions or have direct contact with the firm's funds. Every employee presents a possible hazard.

Losses frequently exceed the amount of insurance carried (Table 3). In some instances the firm's solvency is threatened or destroyed by its loss. No ordinary business concern is normally in a position to carry such a catastrophe hazard itself, and even very large corporations are not justified in self-insuring because, in proportion to the size of their business, the opportunities will be greater for their employees to cause catastrophic losses through dishonest acts.

Various forms of fidelity insurance are available to meet the requirements of any firm whether it is large, medium, or small. To begin with, there is the individual bond. If the firm has only one or two employees, it can purchase for each an individual bond that is issued on behalf of the named employee and for

Figure 4. Exposure Index Related to Coverage
(Courtesy of The Surety Association of America)

Chart A

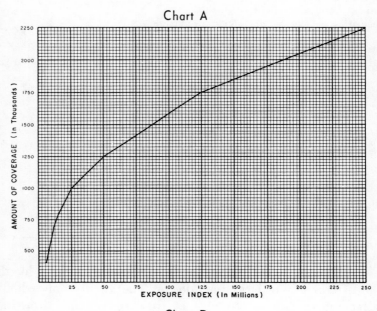

Chart B

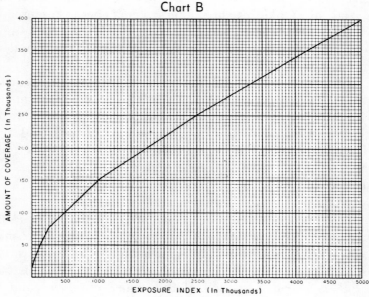

Table 3. Examples of Losses Caused by Dishonest Employees

Business	Employee	Loss	Bond	Un-insured Loss
Wholesale produce	Bookkeeper	$185,820	$25,000	$160,820
Dress manufacturer	Dept. manager	43,000	20,000	23,000
Plumbing supplies	Warehouseman	17,000	2,500	14,500
Retail dairy	Office manager	11,000	2,500	8,500
Furniture dealer	Credit clerk	22,000	5,000	17,000
Coal and Ice	Bookkeeper	28,240	5,000	23,240
Wholesale grocer	Salesman	29,345	12,500	16,845
Hospital	Chief clerk	15,000	5,000	10,000
Paper mill	Paymaster	45,000	10,000	35,000
Public utility	Treasurer	99,139	25,000	74,139
Machinery manufacturer	Sales manager	96,940	50,000	46,940
Export-import	Bookkeeper	65,891	20,000	45,891
Department store	Several	81,000	15,000	66,000
Meat packer	Superintendent	43,404	10,000	33,404
Automobile dealer	Distributor	98,700	50,000	48,700
General merchandise	Cashier	22,000	3,000	19,000
Heater manufacturer	Manager	30,000	10,000	20,000
Refrigerator manufacturer	Cashier	20,810	5,000	15,810
Rubber manufacturer	Bookkeeper	126,700	26,000	100,700
Steamship company	Asst. cashier	65,000	1,000	64,000
Advertising	Billing clerk	90,875	10,000	80,875
Auto dealer	Bookkeeper	31,361	10,000	21,361
Department store	Floor manager	18,500	10,000	8,500
Foundry	Bookkeeper	36,000	7,500	28,500
General merchandise	Manager	15,200	10,000	5,200
Grain dealer	Elevator manager	26,306	10,000	16,306
Hardware	Credit manager	40,871	10,000	30,871
Radio tube manufacturer	Several	48,000	20,000	28,000
Novelty manufacturer	Shipping clerk	34,696	12,500	22,196
Oil drilling	Supervisor	30,882	10,000	20,882
Paper products	Warehouseman	25,551	15,000	10,551
Rubber products	Office manager	150,500	25,500	125,000
Tobacco products	Bookkeeper	43,000	10,000	33,000
Wholesale grocer	Cashier	52,348	10,000	42,348

a stated amount. The bond may not be amended to cover any other employee.

Schedule bonds. If there are several bondable employees, a schedule form of bond would be advisable. There are two types of schedule bonds: name schedule bond and position schedule bond. The name schedule bond covers the employee by name in the amount set opposite his name. That form reduces detail: although it is similar to the individual bond, it covers more than one named employee.

The position schedule bond provides protection against the dishonest acts of the employees who occupy the positions listed on the schedule in the amounts stipulated opposite the positions. By such a bond, the incumbents of designated positions are bonded even though the incumbents are not identified by name in the bond.

If the employer has three different individuals successively occupying the bonded position of, say, messenger during the course of a year and a name schedule bond is carried, a new acceptance specifying the amount of liability assumed must be obtained for each incumbent. Under the position schedule bond, on the other hand, no change would need to be made regardless of the number of individuals who might succeed one another as messenger during the term of the bond.

Blanket bonds. If there are a larger number of employees to be bonded, a blanket fidelity bond is recommended. There are two forms. The commercial blanket bond covers all officers and employees collectively. In the event of a loss, regardless of the number of employees involved, the aggregate amount collectible is the bond penalty. The bond is issued in a minimum penalty of $10,000 and has no maximum or ceiling. The second form, the blanket position bond, also covers all employees, but in the event of a collusive loss the bond penalty applies to each identifiable employee involved in the loss. The bond runs from a minimum penalty of $2,500 to a maximum of $100,000. Both forms of bonds automatically cover all new employees during the term of the bond without notice to the insuring company and without additional premium charge.

If a concern has a large employee exposure or could conceivably be subjected to a catastrophic loss from the dishonest acts of one employee or a group of employees, the commercial blanket bond is preferable in an amount estimated to cover possible losses. If the concern has a relatively restricted exposure and losses may be expected to run within a reasonably limited area, the blanket position bond appears to be the logical selection, particularly in view of its collusive loss feature.

Although a blanket position bond may be sufficient for some firms and a commercial blanket bond for others, a combination of both forms may be arranged for larger firms with greater exposures. When the bond amount under a blanket position bond is $10,000 or over, the larger firm will find advantageous a blanket position bond as primary coverage and a commercial blanket bond as excess leverage. Thus, the blanket position bond provides basic protection against loss caused by employees acting in collusion; and if that proves to be insufficient, the commercial blanket bond provides the excess coverage for the larger loss, whether caused by one or several employees.

When a number of employees are to be bonded, blanket bonds are made to order, since they offer the greatest protection for the premium dollar and compare favorably in premiums with the more limited individual and schedule bonds. Furthermore, the danger of an unbonded loss is minimized, since all eligible employees are covered to the full amount of the bond and there can be no possible failure through oversight to include individual employees under a blanket bond.

All premium rates are on an annual basis, but any fidelity bond may be written on a three-year premium plan. If fully paid in advance, there is a reduced term premium that can effect a substantial saving if the coverage is reasonable enough to begin with.

The principal advantage of the blanket position bond over the commercial blanket bond is its collusive loss feature. Obtainable in this form only, it permits recovery for the full amount of the bond on each employee identified as participating in a collusive loss. For instance, under a $10,000 blanket position bond, if five employees act in collusion to create a $50,000 loss, the full amount is recoverable if all five are identified. If the guilty employees cannot be identified, recovery is still available in the amount of $10,000, or the full penalty of the bond. If, on the other hand, a commercial blanket bond of $10,000 were in force in that case, total recovery would be $10,000 whether or not the employees were identified.

Comprehensive 3-D policy. Although it is not purely a fidelity form, the comprehensive dishonesty, disappearance, and destruction policy (popularly termed the 3-D policy) fills a unique place in the protection afforded the business firm or other organization with an employee payroll, since it covers not only employee dishonesty but other hazards as well. The policy has five basic insuring agreements, and other insuring agreements are available by endorsement. Coverage may be written under any one or more of the five basic agreements, and each is independent of the others as to specific protection and amount of coverage.

Two of the insuring agreements are of special interest to employers who wish protection against employee defalcation. One covers loss of money, securities, and other property caused by the dishonest acts of any officer or employee. It is available under either of two forms to be selected by the employer: form A, which is the commercial blanket bond type of coverage, and form B, or blanket position bond type. Both forms have already been discussed.

The other pertinent agreement covers loss through forgery or alteration of checks, drafts, and other specified instruments issued by the insured. A great many of the very substantial employee dishonesty losses include elements of check forgery, and therefore many concerns purchase that type of forgery coverage as inexpensive and supplemental employee dishonesty coverage.

Blanket crime policy. The blanket crime policy provides coverage quite similar to that of the comprehensive dishonesty, disappearance, and destruction policy with the important difference that a single limit of insurance applies to all its coverages. Unlike the 3-D policy, however, none of the basic agreements may be eliminated at the option of the insured. The blanket crime policy is designed for those who prefer to buy insurance coverages in a single amount rather than select varying amounts. As its name indicates, it is a blanket policy and covers all employees, locations, and messengers of an insured.

Detection

Employees should not be made aware of all of the details of the operation of the detection system, because an air of mystery will act as a psychological deterrent to those who might be motivated to perform acts that would be detrimental. Everyone in the organization should be aware that a detection system is in operation; and if the detection portion of the program is properly explained to them, the employees will accept it as a normal aspect of management control.

Inspections and Spot Checks

Guards as well as other representatives of the security organization should make frequent, unannounced inspections and spot checks of all the operations in the facility. In addition, a self-inspection system will prove valuable as a detection aid at the operating levels. Supervisors throughout the facility can be instructed to supervise inspections of their own areas in an effort to uncover discrepancies.

The inspection system should be designed to determine if security procedures are being followed and if the preventive measures are operating effectively. All elements of the protection plan must be constantly checked and reviewed. That is necessary not only to insure that all aspects of the plan are functioning effectively but because protection requirements change. Security inspections may reveal gaps in the protection program, or they may reveal that security controls are no longer needed in some areas because of changes in the facility or the work being performed.

Testing the System

As a means of testing the protection system and employees, created errors can be inserted into the system. The technique can be utilized to determine whether procedures are being followed and personnel are performing the protection duties assigned them. If the errors are not reported, then the security organization should conduct an immediate thorough investigation to determine the corrective action needed. For example, an arrangement might be made with a supplier to ship more of an item than was ordered to determine if personnel in the receiving area are detecting and reporting discrepancies in quantity and quality received.

Test exercises can be designed and conducted to determine the effectiveness of the emergency plan designed for the facility. Exercises involving the type of emergency situations that would realistically be expected can be written and presented for solution. Deficiencies in the plan, as well as unrealistic features, should be highlighted. Any procedures that require revision should be noted. Those responsible for administering the plan will receive valuable training in reacting to simulated emergency situations.

All aspects of the protection program should be constantly tested so that deficiencies in the system will be revealed and individuals who are causing problems that might result in serious losses can be detected.

REFERENCES

1. Jane Jacobs, *The Death and Life of Great American Cities* (New York: Random House, Inc., 1961). By permission.

2. The information concerning honesty insurance was extracted from material published by The Surety Association of America, 110 William Street, New York, N.Y. 10038

Self Test Questions

1. What are the three essential elements that should be included in a protection program if it is to be successful?

2. What are the three groups in an enterprise that should be included in an education program?

3. Why is an education program considered the cornerstone of any effective assets protection program?

4. What is one important preventive measure that should precede the education program?

5. Should insurance ever be utilized to neutralize all risks an enterprise may face? If not, why not?

6. List two measures that may be utilized in the implementation of a detection system.

$$=== 4 ===$$

Prevention of
Industrial Espionage

INCREASING business competition in recent years has resulted in constantly expanding needs for new inventions and products. Consequently, expenditures by American industry for research and development have increased from $90 million in 1920 to more than $54 billion in 1979. It was estimated that $23.8 billion had been spent in the private sector and that $31 billion had been spent by the United States government.[1]

It is not enough simply to exploit inventions and improvements that result in better products and processes; the information involved must be protected. In order to realize a profitable return on the large investments now being made, companies must be in a position to successfully prohibit competitors from using the results of research and development without permission. The prevention that was stressed in Chapter 3 is the key to that program.

Top management representatives may feel they do not have any information to protect because they may not be involved in research and development efforts and are not developing a large number of new products. They may feel that they do not have trade secrets worth stealing. However, the potential value of information in every organization is indicated by the following quotation:

> It is as mundane as a price change or as exotic as Coca-Cola's seventh ingredient. It is as simple as last month's performance or as complex as a trade secret; it is as commonplace as a firm's plans or as rare as a new patentable process. It covers everything your company has that could be of benefit to your competitor.[2]

Secrecy has always been regarded as the best method of protecting the results of intellectual effort. To maintain secrecy, rulers in ancient times often had architects and engineers killed after they had completed their work. Medieval guilds and craftsmen in preindustrial Europe and America imposed severe restrictions on apprentices and their future activities. However, many members of modern society do not regard secrecy as being in the public interest because they feel that the widest dissemination of new advances in technology and culture will benefit the public.

Legal Protection

Patent and copyright laws were enacted in both England and the United States at about the time of the industrial revolution to encourage inventors and authors to disclose their efforts. The objective of the legislation was to encourage the progress of science and the arts by giving inventors and authors limited monopolies. In the United States, a patent grants an inventor or his assignee the right to exclude others from making, using, or selling the invention in the United States for a period of 17 years. Anyone who desires to use the invention during that period must obtain permission from the inventor or his assignee in the form of a license and pay a royalty. The invention becomes public property at the end of the 17 years. To be patentable, an invention must be a new and useful process, a machine, a manufactured product, a composition of matter, or a design.

Although the copyright and patent legislation attempted to stimulate creativity, there is evidence that the systems themselves have failed to meet changing technology and have not been able to cope with changing social and industrial practices. An excellent example is the failure of both patent and copyright provisions to provide protection for computer software. Also, the 17-year period of protection is often regarded as inadequate because inventions may require a much longer period to be profitable after they have been reduced to commercially acceptable products. In addition, a patent issued in the United States can be enforced only in the United States, so it is necessary to apply for a separate patent in each country where there is a threat of infringement. The disclosure of an invention may also serve to inform a competitor of a new state of the art and allow him to design around the patent. Patent information is very easy to obtain. A copy of any patent can be purchased from the U.S. Patent Office for 50 cents.

Because of the deficiencies of the patent and copyright system, many U.S. companies prefer to protect the confidential nature of inventions by establishing them as trade secrets. There is no limit on the life of a trade secret. Also, a discovery may not meet the exacting requirements of the Patent Office, which refuses to issue about one-fifth of the patents applied for. If an invention that cannot be patented is to be protected, it must be treated as a trade secret. According to Restatement of Torts, Section 757:

> [A trade secret is] any formula, pattern, device or compilation of information which is used in one's business and which gives one an opportunity to obtain an advantage over competitors who do not know and use it. It is a process or device for continuous use in the operation of the business. Generally, it relates to the production of goods as, for example, a machine or formula for the production of an article.

The subject matter of a trade secret must be something of value that requires considerable effort and expense to develop. It should be information that a competitor would not be able to learn except by the use of improper means. The use of spying and bribery to obtain a trade secret has been held to be improper means.

Although considerable legislation has been passed to protect copyrights and inventions, there has been a serious lack of statutory protection designed for the protection of trade secrets. However, throughout common law it is recognized that trade secrets must be protected. Some items that the common law recognizes may be protected as trade secrets are descriptions of inventions, plant layouts, shop know-how, formulas for new processes, methods of quality control, customer lists, and marketing data. To qualify as a trade secret, no proof of invention or originality is required; it is necessary that the information is not known to the general public or to the trade. Also, of course, it must be of value to the organization possessing the information. The owner of a trade secret may communicate it to employees involved in its use without losing protection.

Courts in the United States have held that two basic requirements must be met if legal protection for a trade secret is expected. The techniques of meeting those requirements will be discussed later. The requirements are: (1) If a trade secret is to be disclosed to another party, notice must be given that it is a trade secret along with notice that there is a prohibition against further disclosure. (2) The trade secret must be capable of protection and the owner must take positive steps to protect and preserve its integrity.

Company representatives who have had experience with cases involving the loss of intellectual property advise that adequate legal redress can usually never be obtained after a trade secret has been lost or stolen. Two factors are involved: the high cost of litigation and the difficulty of obtaining a judgment for the amount of the loss. Even if a judgment has been obtained, it is often impossible to collect anything. A classic example, one of the most famous United States cases concerning trade secrets, involved a Dr. Robert S. Aries.

Merck & Company had developed a formula at a cost of $1.5 million and it alleged in a suit that Dr. Aries had put the formula to an unlawful and personally profitable use. A judgment was obtained against him by the company for more than $6 million after many thousands of dollars were spent on litigation. However, Merck has never been able to collect because Dr. Aries is reported to be living in luxury on the Riviera in France. Although Merck has a judgment in the form of a piece of paper proving that it is owed more than $6 million, the value of the judgment is only the value of the paper on which the judgment is written.

Civil remedies only have been considered up to this point. The criminal aspect of the improper obtaining of trade secrets should also be mentioned. Until recently, many jurisdictions have not provided for statutory law to cover such thefts. Trade secret thefts in most jurisdictions have had to be treated under tangible-property laws. It was extremely difficult to make a case for an indictment under property laws when only data were involved. If papers were stolen, only the value of the paper could be claimed—not the value of the data contained therein.

One criminal case that was prosecuted in federal court, *United States* v. *Mayfield,* involved theft of a marketing plan. It was decided on August 5, 1965, in the criminal division of the U.S. District Court, New York Eastern District. Eugene Mayfield, a few months after he had resigned from a junior executive post at Procter & Gamble Company in the summer of 1964, telephoned a man at Colgate-Palmolive Company in New York City. During the conversation, which he carried on from his home in Illinois, Mr. Mayfield offered to sell to Colgate for $20,000 P&G's plan for marketing its Crest toothpaste during the 1964–65 fiscal year. Colgate immediately reported the matter to the FBI.

The Colgate man, acting under instructions from the FBI, continued negotiating for the purchase of the stolen trade secret. He agreed to meet Mr. Mayfield in a men's washroom at the Kennedy International Airport in New York City. At the time arranged, he entered a compartment of the washroom and was directed by Mr. Mayfield, who occupied the adjacent compartment, to hand the $20,000 under the partition, along with his pants. The device intended to prevent pursuit was not effective, because when Mr. Mayfield dashed out of the washroom with the money, FBI agents were waiting to arrest him.

The incident involved two federal statutes. One makes it a crime to use the telephone in interstate commerce for fraudulent purposes. The second makes it a crime to carry stolen goods across state lines. Mr. Mayfield was charged with violating the second statute. During the trial, it was pointed out by the prose-

cuting attorney that the stolen marketing plan was a trade secret
that P&G valued at more than $1 million. Mr. Mayfield pleaded
guilty to the charge and was sentenced to two years' imprisonment, but the sentence was suspended. The maximum penalty
for transporting stolen goods across state lines is ten years' imprisonment and a fine of $10,000.

Negligent and Willful Disclosure

Any plan designed to prevent loss of trade secrets must take
into consideration both negligent and willful disclosure. Negligent disclosure is usually made by an employee. It may involve
only a portion, although it could be the most important portion,
of the trade secret. The conclusion that the individual "meant
well," "simply talked too much," or "did not know" will be
of little comfort if the information disclosed has an adverse impact on the profits of the organization. Almost any employee
who has knowledge of a trade secret can negligently disclose
it. That type of negligence may vary from the careless handling
or storing of papers to the boasting of a salesman or other employee to the wrong person or at the wrong place.

The willful disclosure is more serious because it is more difficult to control. Also, the disclosure may involve all of the data
involved in the trade secret at one time or over a period of time
until a competitor has all the valuable information he needs.
The possible result is that the organization attempting to safeguard the secret will be seriously injured. Sources of collection
may involve individuals, records and materials, products, surveillance, pretexts or false acts, bribery or blackmail, placing an
espionage agent in a competitor's facility, or theft of information
or material. Willful disclosure can be divided into two general
categories: open and hidden.

Open and Hidden Collection

Individuals who openly collect information may be employees
of the organization that is attempting to safeguard material, com-

petitor's employees, exemployees, or representatives of other
organizations as well as vendors, suppliers, or contract service
representatives.

Salesmen and company field representatives may supply trade
secrets to those who are openly collecting information. Such employees are constantly dealing with a large variety of outside
contacts. In addition, they are often extroverts by nature and
like to talk to people. They may even be anxious to impress
their contacts with the importance of their positions.

Another class of employee who deals with the public and who may
possess sensitive information is the purchasing agent or buyer.
Buyers may be required to have trade secrets in their possession in
order to obtain the best material at the most favorable price. The use
of that sensitive information and how much they transmit to those
who whom they are dealing may create a problem.

Attendees at conferences, seminars, or trade shows may be
an excellent source for the individual who is openly collecting
information. Those who attend meetings often feel they are required to "give a little to get a lot" of information from the
competitors. The question, of course, is whether they really get
more than they give.

Consultants should not be overlooked. A consultant may be providing services to a number of organizations simultaneously, and
may be an excellent, although quite unwitting, source of information for a representative of a competitor. Few, if any, consultants
would provide information to a competitor as a part of a consulting
contract.

Another source for open collection of information is the employment interview and applicant résumé. A competitor may use
this technique to good advantage with an individual who would
like to change employment or is actually looking for a position.
License and merger negotiations may also be utilized to obtain
information that might not otherwise be released, especially
from an organization anxious to complete a license arrangement
or effect a merger.

Other sources for the individual or organization that is
openly collecting information are vendors, suppliers, and con-

tract service representatives. For example, an unethical contract janitorial service or an individual worker of the service could be utilized effectively. Free access to the president's office in most organizations is carefully controlled, but janitors from an outside organization may be there on a regular basis after the close of business. They are usually there alone with ample time to collect information if that is their objective.

Published materials are an excellent source for those who are openly collecting information; they include scientific and technical papers in professional journals, magazine and newspaper articles, surveys and reports, and sales brochures. Unpublished materials include records of government agency hearings, speeches at panels and conferences, court records, and patent files.

Hidden collection methods are also effective in obtaining trade secret information. One technique is a surveillance of facilities or individuals. That may include visual observation, with or without trespass, or electronic observation.

Pretext interviews may be used with employees of a competitor when the intent is not to hire but only to obtain information. Another technique that utilizes employees of a competitor is to hire them only with an intent to gain the know-how of the competitor.

Bribery or blackmail of individuals may also prove effective in obtaining sensitive information. Still another method is to place an espionage agent or plant in a competitor's facility. That could be an employee of the competitor, a service representative working in the competitor's plant, or a guest or visitor to the facility of the competitor.

A more brazen method of collecting information is to simply steal it—a technique that has been utilized frequently.

The Mobile Employee

The mobile employee in the United States has caused a great deal of concern because this type of individual has made the safeguarding of trade secrets more difficult. Scientists and executives feel they should be able to move from job to job without sacrificing either personal or financial status. A celebrated United States case popularly known as the Goodrich or space suit case involved the problem of an employee going to work for a competing company after being given trade secrets by the original employer. Two basic issues that may be in conflict were involved in that case. The first issue is the right of an organization to its intellectural property or trade secrets; the second is the right to seek gainful employment in which the individual's ability and experience may be utilized.

Trade secrets that companies may attempt to protect are often an important aspect of a departing employee's total capabilities. Intellectual capacity may be a combination of information acquired from the individual's employer,   co-workers, and the individual's own research efforts. Such information, if treated by the company as a trade secret, can be regarded as corporate property because the normal employment contract or other employment arrangement usually specifies that all employee-generated data, including inventions, become the property of the company. As a result, a significant amount of a departing employee's intellectual capacity can be claimed to be corporate property.

In a new position with a competitor the trade secrets in the possession of a former employee will begin to be utilized, often subconsciously. For example, the individual will use his overall intellectual capacity, which will often include the trade secrets of his former employer, in making daily decisions or in developing plans for future projects.

Contractual limitations have been utilized to protect an organization against possible damage by the activities of the mobile employee. The prohibitions are of two general types. One limits the disclosure and use of trade secrets by employees during their employment or at any time thereafter. The other provides for restraints against certain future activities of employees following their employment.

A restraint against unauthorized disclosure or use is normally upheld in the courts provided it is limited to legitimate company-defined trade secrets. However, owing to unintentional release of information and subconscious utilization of trade secrets, it is often difficult to prove that a violation has occurred. Unless there is also a restraint against future employment with competitors, this type of an agreement may be ineffective. Limiting future activities may be difficult to define contractually. Some agreements in the past limited certain activities of the departing employee for a period of time and within a geographical sector. Because company interests are often national in scope, a geographical limitation may be meaningless.

Courts have been reluctant to limit the activities of an individual when his freedom to use his skills are at stake. Companies also face the difficulty of convincing a court that a limitation on future employment is necessary. The court will usually look for evidence that the former employee has or will, in fact, be clearly utilizing more information than an ordinary employee of his competence could be expected to use. Some states prohibit future-employment restraints. It would now seem that a broad contractual limitation on employment that would prevent individuals from using their overall capability will usually be held invalid.

Prevention Program

In setting up a program to prevent industrial espionage, a number of steps should be taken:

- Information to be protected should be defined and provisions should be made for marking it.
- Transmission, handling, storage, and destruction procedures should be defined.
- Policy regarding release of information should be determined.
- Written procedures should be published.
- Terms of agreements to be executed between the company and the employee handling sensitive company information should be defined.
- Employees should be educated.
- Disciplinary actions to be taken in case procedures are violated should be defined.

Definition and Marking of Material

Information that falls into the trade secret category must be
defined if it is to be safeguarded. If it is not defined, large
amounts of material that is really not in the trade secret class will
be included and that will hamper the efficient operation of the
system. Selectivity in determining what is to be protected is essential. Money will be saved because only the material that requires
protection will be included in the protection plan, and employees
will more readily accept the program because they will understand the need to protect the selected information.

After the information to be secured has been selected, it
should be marked with an appropriate designation that will make
everyone handling it aware that it is to be protected. Earlier
in this chapter, under "Legal Protection," it was pointed out
that one of the basic requirements for protection under the law
is that notice has to be given when information is regarded as
a trade secret. Marking the material will help meet that requirement for legal protection. A variety of designations may be used,
but in the United States, because the Department of Defense
security program uses them, many organizations have avoided
top secret, secret, and confidential. Instead, such designations
as company private, company proprietary data, and internal data
have been used.

Transmission, Handling, Storage, and Destruction

A most important aspect of trade secret information processing
is the establishment of a need-to-know practice. Only those who
have a need to know the information should receive it. The individual who generates or transmits the information must assume
the responsibility for determining those who have a need for it.

The proper safeguarding of intellectual data is particularly
important during transmission, handling, storage, and destruction. Much of that type of information can go astray if it is not
properly handled. Complex trade secret information that might
have cost hundreds of thousands of dollars to develop may be

summarized and reduced to a short memorandum for the use of key officials of the organization. Practices should be established for everyone handling such material. The procedures need not be complicated or complex, but they should furnish control that will prevent disclosure of sensitive information.

Usually, trade secret information is transmitted within the organization in sealed envelopes that have been marked to indicate the sensitivity of the information. The envelopes are generally given special handling within the mail system. A document control system may be used to insure that all copies are accounted for and that each copy is received by the individual to whom it is addressed. Also, trade secret information is generally locked up when it is not in use—at least in a key-locked container.

Destruction of material after it has served its useful purpose is also important. Espionage agents have always considered trash an excellent source of information, and that should be considered when papers or documents that contain trade secret information are to be disposed of. Burning is always an effective method of destruction.

Release of Information

A policy concerning the release of information should also be established. It should be designed not only to regulate the information released by the public relations department as a part of its daily activities but also to control information that may be released by all employees. Employees at almost every level belong to professional societies that are constantly trying to motivate individuals to present papers at meetings or publish material dealing with their fields of specialization. That is particularly true of engineers and scientists who may be working with sensitive company technical information.

Unless a review procedure has been set up to screen all information to be published or presented at meetings, trade secrets may unwittingly be included in the presentations. Written approval in most organizations is required before material can be released by employees.

Publication of Written Procedures

All procedures established for the protection of trade secret information should be reduced to writing and published so that those who handle such information will be aware of the responsibility they must assume. Many organizations have manuals that outline not only procedures to be followed in the case of trade secrets but also practices to be followed with respect to inventions and patents.

Agreements Between Employees and Company

Some of the general legal aspects of contractual restrictions were discussed under the heading "The Mobile Employee" earlier in this chapter. Since employees are often required to sign an agreement dealing with patent and trade secrets restrictions as a condition of employment, some of the aspects of actual agreements will be outlined in this section.

Which employees will be required to sign agreements varies from company to company. One organization may feel that adequate protection for investments in research is secured if only research employees and supervisors sign the agreements. Other employers may feel it is necessary to include salesmen; still others may have conditions such that all employees will be included.

The agreements in use vary in length, format, mode of expression, and scope of coverage. The National Industrial Conference Board made a survey of 83 companies in 1965 and found that the forms varied in length from one page to eight pages. Some company executives are of the opinion that a company may stifle the creative thought of employees when it includes too many restrictions in patent and secrecy agreements. Agreements that are long and complicated may arouse suspicion and endanger goodwill.

Many of the agreements in use in the United States have the appearance of a standard legal contract. Other companies may use a memorandum or letter that is sent to the employee over the signature of a company official. Employees sign the memorandum or letter and returns it to the company if they agree to the terms and conditions outlined. In some companies, the agreement is made an integral part of the employment form.

Some companies have incorporated a provision in patent and secrecy agreements that specifically prohibits moonlighting by employees. The companies feel that holding a second job can put such a strain on employees that their efficiency on the job may be diminished. They also feel that the second employer may be a competitor who hired the employee in order to acquire the trade secrets of the full-time employer.

Education

As was pointed out in Chapter 3, employee education is one of the key elements in any security program. It is also an essential element in an industrial espionage prevention program, because the cooperation and support of employees is required. Disinterested or apathetic employees will have a really detrimental effect on the best plan. Often such attitudes result from a lack of understanding of the need for the program.

Publication of procedures and the distribution of a manual has already been mentioned under "Publication of Written Procedures." Indoctrination lectures and discussions in staff meetings are also effective techniques. Whatever methods are used to inform employees of the details of the plan, it is important that the presentation be made in a serious manner. Posters may be placed throughout the facility to give employees instructions concerning the program. The practical and realistic aspects must be emphasized. It is necessary to make industrial espionage prevention easy to live with as part of the normal work routine.

The selection of the proper media requires careful planning; the best instruction material is of little value if it is not understood. Many different methods are available to tell employees about the company's program. The lines of communication selected should be those that seem to be the most sought after by the group to be reached. Advertising and public relations experts in the organization may be of considerable help in preparing effective copy and in selecting the best media.

In searching for the proper media, the first step should be

an appraisal of the established information pipelines in the company. Those ways of disseminating information should be effective, since employees are usually already familiar with them and naturally turn to them.

New employees should not be overlooked. The sooner they become acquainted with the program the better, because they will probably take more of an interest in it if it is explained as an important company activity early in their employment.

A security debriefing should not be overlooked when an employee terminates. A short meeting with the employee before will give a management representative an opportunity to discuss restrictions on the employees's future activities and caution the individual about revealing trade secrets that might have been obtained in the course of employment. Some companies utilize a security termination statement form in connection with the debriefing. Employees are asked to sign the form, in which they certify that they understand the limitations on future activities with respect to trade secret information.

Discipline

Even if all the safeguards are implemented, employees will violate the established practices wilfully because of carelessness, indifference, or lack of understanding. Therefore, a system of correction must also be designed so that the proper administrative or disciplinary action is taken in each instance. Employees should be aware that a disciplinary program does exist and that action will be taken when there are infractions of the rules. All such actions, as was pointed out in Chapter 3, should be taken within the supervisory chain.

REFERENCES

1. *More Speed Behind R & D Spending,* Business Week, July 7, 1980, page 47 and *Federal Funds For Research and Development Fiscal Years, 1978, 1979, and 1980,* Volume XXVIII, National Science Foundation, page 12.

2. Floyd Purvis, manager of corporate security, Texas Instruments Incorporated, as quoted in *The Great Game of Corporate Espionage, Dun's,* October 1970, page 30.

Self Test Questions

1. Why do many organizations in the United States prefer not to use the patent and copyright system for the protection of inventions but would rather treat them as trade secrets?

2. According to the Restatement of Torts, what is a trade secret?

3. What are two requirements that have been established by courts in the United States that must be met if legal protection for trade secret information can be expected?

4. Why should trade secret information be destroyed after it has served its purpose? How should it be destroyed?

5. Why is it important that an information release policy be established?

6. What should be discussed with a terminating employee who has had access to trade secret information when the employee is being debriefed?

Disorders
and Terrorism

A UNITED States Government report on disorders and terrorism issued in late 1976 stated, "Disorders and terrorism have both common characteristics and specific differences. Both are forms of extraordinary violence that disrupt the civil peace; both originate in some form of social excitement, discontent, and unrest; both can engender massive fear in the community. Disorders and terrorism constitute— in varying forms and degree—violent attacks upon the established order of society."[1]

The report predicts that future terrorist activity can be anticipated in the following quote: "A majority of experts predicts an eventual increase in terrorist activity and an escalation of its intensity." With reference to future civil disorder, the report states, "It is important to remember that some of the most serious episodes of mass violence in this country have been touched off by relatively insignificant incidents. The mood of the community should be monitored constantly; signs of impending violence never should be ignored. Contingency planning should be predicted on the assumption that a deep well of violence underlies the apparent calm and stability of the American social scene."

The warnings contained in this report, together with predictions issued by the Federal Bureau of Investigation and other organizations as well as a 1978 United States Senate report,[2] clearly indicate that disorders and terrorism can be considered major future problem areas that should be given consideration in any security program. As a result, this chapter is being devoted to a discussion of these potentially serious problem areas.

Disorders

Recorded history clearly illustrates that violence resulting from disorders have been a form of protest in all civilizations. Class strife, which was characterized by rioting, is credited with finally destroying the Greek city-states. Riotous mobs of slaves were common in the Roman Empire; the common people and the ruling class at that time were engaged in a continuous social struggle. In the Middle Ages, peasant riots occurred frequently.

During the industrial revolution in the eighteenth century, rioters directed their hostility toward machines as well as the upper-middle-class machine owners, whom they blamed for the reduction in the number of jobs. Whereas the participants had previously been peasants, now they were city dwellers. In 1780, at the time of the Gordon anti-Catholic riots, London was in the hands of rioters for a week.

Serious rioting developed in Manchester in 1819, when 50,000 people gathered to insist on parliamentary reform. Numerous casualties were inflicted when the cavalry charged the gathering. The result is known as the Manchester Massacre. There were violent riots in Baltimore in 1812, in Boston in 1837, in Philadelphia in 1838, and in Chicago in 1886.

The depression that started in 1929 introduced a period of economic disturbances and strikes. Riots developed in the Midwest as the result of demonstrations against mortgage foreclosures.

Racial riots occurred during World War II; two of the most serious developed in Detroit, Michigan, and in New York City during the summer of 1943. The political, social, and economic instability that followed World War II resulted in civil disturbances becoming a rather common phenomenon of the postwar period.

The Supreme Court decision on school desegregation in 1954 and subsequent decisions resulted in racial turbulence and demonstrations in both the southern and northern parts of the United States. In 1964, riots developed in seven northern cities. In the summer of 1965, business and industrial activity came to a shocking standstill across a 50-mile swath in the southern part of Los Angeles during the Watts riots. When the riots were finally brought under control after six days, 34 persons were dead, 1,032 were wounded or hurt, and property damage was estimated at $40 million. Also, 232 business establishments were completely destroyed, 632 were damaged and 41 buildings were completely wiped out.

The worst riots in the United States occurred during the summer of 1967. About 50 serious incidents of crucial civil disorder developed throughout the country that year. Property damage was estimated at about $160.4 million; deaths were estimated at 83; about 1,953 were injured; and more than 16,471 arrests were made. On July 12, 1967, the most serious outbreak since the Watts riots in 1965 took place in Newark, New Jersey. After six days, 26 were dead, 1,500 were injured and an estimated $30 million in property damage had occurred. That was followed by an even more grim outbreak in Detroit, Michigan, during the period July 23 through July 30, 1967, when 43 died, 2,000 were injured and property damage was an estimated $45 million.

In April 1968, after the Rev. Martin Luther King, Jr. was killed in Memphis, Tennessee, a riot developed in Washington, D.C. which lasted for four days. Nine died and 1,000 were injured. The violence extended to other cities—particularly Baltimore, Chicago, Detroit, Memphis and Pittsburgh.

These were not race riots involving whites versus blacks such as those that occurred in Detroit and New York City during World War II. Instead, they appeared to be hostile outbursts of segments of the population of black communities located in large northern cities. The 1970s went relatively without disorders as the country seemed to enjoy a decade of racial peace.

However, as if to verify the predictions made in the 1976 U.S. Government report quoted earlier in this chapter, a disorder again took place early in 1980 in Miami, Florida. In late May, violence erupted there and after three days of rioting, 17 were dead, about 400 were injured and damage was estimated at $200 million.

Although social protest disturbances have been highlighted in recent years, that is not the only kind of civil disturbance. In general, there are five kinds: social, economic, political, disaster, and absence or failure of constituted authority.

Social disturbances result from racial or religious differences or from excitement stemming from a celebration, a sports event, or other social activity. Economic disturbances arise from disagreements during labor disputes and strikes or from such extreme conditions of privations or poverty that the people will resort to violence to obtain the necessities of life. Political disturbances originate with attemps to gain political power by other than lawful means. The conditions following disasters may generate violent disturbances among people because of the fear of further catastrophic effects: lack of food, clothing, or shelter or action of lawless elements. The absence of authority or inability or failure of authorities to exercise their responsibilities may result in a disturbance because persons believe they can violate the law with impunity.

Group Behavior

To cope with civil disorders successfully, an understanding of group behavior is necessary; it must include typical actions and motives, some basic patterns of group behavior, and the individual characteristics that underlie the behavior. Although group activity sets the scene for civil disturbances, a crowd or mob is composed of individuals.

In general, the difference between a crowd and a mob is that a crowd is law-abiding and a mob takes the law into its own hands. A crowd may be defined as a large number of persons temporarily congregated. Generally, the members of a crowd think and act as individuals and are without organization.

Under normal conditions when a crowd is orderly, violates no laws, and causes no danger to life or property, it does not present a significant problem. In the environment of civil unrest, however, any crowd represents a potential threat to the maintenance of law and order. Although innocent in its origin, nature, or purpose, a crowd can develop into a violent group that might and often does pursue a course that ignores law and order. That type of group has the potential of generating a riot.

A mob can be defined as a crowd whose members, under the stimulus of intense excitement or agitation, may lose their sense of reason and respect for law and follow leaders in lawless acts. Mob behavior is essentially emotional and without reason. The momentum generated by mob activities has a tendency to reduce the behavior of the total group to that of its worst members. In mob activities, the first persons who take definitive action usually are the most impulsive, the most suggestible, the least self-controlled, and the least inhibited. The most ignorant and most excitable are the ones who are likely to trigger the violence. Once the violence has begun, it usually spreads quickly and engulfs the more intelligent and self-controlled.

Mobs do not always develop from crowd formations. In today's society, skillful agitators, by making use of radio, television, and other communications media, can reach large portions of the population and incite them to rebellious action without coming into personal contact with them. When agitators succeed in arousing the emotions of persons, whether in a group or individually, they can be expected to act together in either large or small dispersed groups and participate in acts of lawlessness such as sniping, looting, and burning.

Mobs

Regardless of whether violence is the result of spontaneous reactions or is deliberately incited, riotous actions of mobs can be extremely destructive. Such actions may consist of indiscriminate looting and burning or open attacks on officials, buildings and innocent passersby. Participants are limited in their actions only by their ingenuity, the training of their leaders, and the weapons, supplies, equipment, and materials available to them. Although the degree of violence will depend upon a number of factors, such as the type and number of people involved, location, cause of the disturbance, and weapons available, certain types of violence can be anticipated. Some of the typical actions that can be expected from a mob are outlined in the following paragraphs.

Verbal abuse. Verbal abuse in the form of obscene remarks, taunts, ridicule and jeers can be expected. The purpose of this tactic is to anger and demoralize the persons opposing the mob and cause them to take actions that later may be exploited as acts of brutality.

Attacks on personnel and vehicles. Groups of rioters can be expected to give vent to their emotions upon individuals and formations. Individuals may be beaten, injured or killed. Vehicles may be overturned, set on fire, have their tires slashed, or be otherwise damaged. This type of violence may be directed against personnel and equipment of fire departments and other public utilities.

Thrown objects. Objects may be thrown from various vantage points, such as windows and roofs of nearby buildings and overpasses. The objects may include rotten vegetables and fruits, rocks, bricks, bottles, improvised bombs, Molotov cocktails or any other objects available at the scene.

Vehicles or objects. If those protecting the facility are located on a slope or at the bottom of a slope, dangerous objects can be directed at them such as vehicles, trolley cars, carts, barrels, and liquids. On level ground, wheeled vehicles can be driven toward them and the drivers can jump out before the vehicles reach the target. This tactic may also be used for breaching roadblocks and barricades.

Use of fire. Rioters may set fire to buildings and motor vehicles to block access roads or create confusion. They may flood an area with gasoline or oil and ignite it or pour gasoline or oil down a slope or drop it from buildings and ignite it.

Demolitions. Natural gas, dynamite, or other explosives may be employed. Charges may be placed in the streets or in buildings or be employed to breach a dike, levee or dam to flood an area. Explosives may also be utilized to demolish an overhead bridge or to block a road or street. Dogs or other animals with explosives attached to their bodies may be driven ahead of the mob and the charges then exploded by remote control, fuses, or a time device. Vehicles may be filled with explosives and rolled or driven into a facility.

Use of firearms. Leaders may direct that weapons be fired to encourage the mob to more daring and violent action. That may take the form of sniping or a heavy volume of fire from buildings or from the mob.

Looting. When buildings are broken into, the individuals in the mob will begin to loot.

Other actions. Mob leaders may place women and children or wounded war veterans in the front rank to play on the sympathy of those opposing them. Rioters may take photographs of individuals or groups in an effort to embarrass them. The mob can attach grapples, chains, wires, or ropes to barriers and pull them down. Vehicles can can be crashed against barriers to breach them.

Psychological Factors

There are a number of psychological factors that may influence mob behavior. Some of them are anonymity, impersonality, suggestibility, contagion, release from repressed emotions, novelty, imitation and numbers.

Anonymity. Mobs are anonymous, both because they are large and because they are temporary. The size of the group and the nature of the interaction remove the sense of individuality from the members. The members do no pay attention to other members as individuals and do not feel that they themselves are being singled out as individuals. Thus, the restraints are reduced and an individual feels free to indulge in behavior that he would ordinarily control or avoid, because moral responsibility has been shifted from him to the group.

Impersonality. Group behavior is typically impersonal: the soldier bears no personal grudge against the particular enemy soldier he shoots; in college football, it does not matter that an opposing player is a personal friend. The impersonality of mob behavior is revealed in race riots, when one member of either race is as good or bad as another. When interaction becomes personal, it changes from group to individual behavior, and the nature of the action differs.

Suggestibility. Mob situations are normally unstructured; individual responsibility has been shifted to the group. The situation itself is often confused and chaotic. In such a state of affairs, people act readily and uncritically without conscious realization and without raising rational thought or objections.

Contagion. The most dramatic feature of mob behavior is the emotional buildup which members give to one another. This communication of feeling is most impressive in riots. The members of a mob stimulate and respond to one another and thereby increase their emotional intensity and responsiveness. The process helps to explain why mob behavior sometimes goes farther than some of the members intended.

Release from repressed emotions. The prejudices and unsatisfied desires of the individual that are normally held in restraint are readily released in a mob. The temporary release is a powerful incentive for individuals to participate in mob action, because it gives an opportunity to do things that they wanted to do but did not dare to do.

Novelty. Novelty is demonstrated by the individual who, when confronted by new and strange circumstances, does not respond according to an usual pattern of action. The specific stimuli that usually govern the individual's actions may be absent. The lessons of previous experiences that were employed in solving customary problems may not be applied. The individual may even subconsciously welcome the break in his normal routine and react enthusiastically to new circumstances.

Imitation and numbers. Imitation is the urge to do what others are doing, and the power of numbers may give some members of the mob a feeling of strength and security.

Civil Disturbance Phases

A civil disturbance, from the time it begins until it ends, has been divided into four phases. The precrowd phase is the preparatory time prior to the gathering of a crowd when agitation is below the surface and no precipitating event that would result in the congregating of a crowd has occurred. The crowd phase is the time when, as the result of events, grievances, or agitation, a crowd has gathered. The civil disturbance or riot phase is the time when the crowd has developed into an unruly mob and social disorder prevails. The post civil disturbance phase begins when social order has been restored.

The riot or civil disturbance phase can be divided into three stages. The first stage is generally nasty and brutish. Police are stoned, crowds collect and tension mounts. The second stage is reached with the breaking of windows. Local social control breaks down and the population recognizes that a temporary opportunity for looting is available. If the mob is not dispersed and order restored, the third stage may begin. It is characterized by arson, fire bombing, sniper fire, and countermeasures taken by police and soldiers.

Advance Preparation

The key to dealing with a civil disturbance is advance planning. It must be realized that an individual organization can do little to stop a serious community-wide disturbance and, in most instances, would probably be ineffective in attempting to prevent one from developing.

Policy Decisions

As a first step in the planning process, some basic policy decisions are necessary; some of them involve arrests, prosecution, injunctive relief, employee relations and public relations. The top official or group in the enterprise responsible for policy decisions should act on each item listed so that the position of the organization on each is known in advance. Everyone in the organization who might be affected should then be informed of the policy decisions made.

The question of arrest and prosecution can be decided in advance of a disturbance. With regard to arrests, there are two general types: citizen's arrests and arrests made by local authorities. If arrests are to be made, it must be determined when the local authorities will act and when citizen's arrests should be made, if ever. As a result, it will be necessary to discuss this item with the local authorities and obtain their reaction. If citizen's arrests are to be made, it will be necessary to define who in the organization will make arrests and then instruct them in the proper handling of the arrests. It will next be necessary to decide if those arrested are to be prosecuted. Most local authorities will ask that question when discussing their role, and they will generally expect that officials of the organization will take the necessary tions to insure that those arrested are prosecuted.

An injunction can be used as an important tool to discourage destructive or obstructionist activities of individuals or groups involved in a disturbance. However, preparations should be made in advance of any problem to take advantage of the technique; in particular, steps should be taken to determine the policy position of the organization with regard to injunctive relief. In addition, the requirements for obtaining injunctions should be determined during the advance-planning stage. As much of the detailed paper work as possible should be prepared so that, in case of a problem, it can be quickly submitted to the proper court jurisdiction and a court order can be obtained speedily. Because of the legal factors involved in arrests, prosecutions, and injunctions, it is essential that the legal counsel of the organization participate fully in the policy delibera- tions.

Another important policy decision has to do with action to be taken in the event employees decide to become involved in a disturb- ance. Such involvement might result in employees either joining those causing the disturbance or taking aggressive action on their own against the demonstrators. In either case, a policy decision concerning the limitations the organization will impose on such activities should be reached in advance so that it can be transmitted to employees if necessary.

Policy with regard to public relations should also be determined. For example, will the organization have an aggressive public rela- tions program to keep the public informed, or will no information be released? Since any disturbance is a newsworthy event, the handling of information for the news media and the public should be carefully considered in advance. Any plan made to cope with a civil disturb- ance should include that aspect of the problem.

Planning Factors

The basic requirments for a civil disorder plan are essentially the same as those for any other type of emergency. Therefore, only the items peculiar to the subject of civil disturbances will be considered in this section.

The local law enforcement officials should be contacted as a first step in developing a plan. The type of support that can be expected should be discussed as well as the plan the officials have designed for coping with a disturbance. Such items as the location of the command post to be established in the area and traffic control should be defined. Rest-room facilities, hot food and drink, a rest area for those not actively involved with the demonstrators will be of value to the local officials, and thought should therefore be given to providing them.

Since the demonstrators may want to meet with officials of the organization and make demands, that factor should be considered in the planning process. One individual in the enterprise should be designated to interface with those making the demands. Past experience with demands made by dissidents indicate that some of the demands are often completely unreasonable and some are completely realistic and possible to honor. It has been recommended by those experienced with negotiations with dissidents that a top official of an organization not meet with demonstrators. Instead, a responsible official who will not be empowered to make any decisions on the spot or act on any of the demands should be designated. The interfacing official should be required to receive all demands from demonstrators and transmit them to the top officials for action. He can then give the reply to the demonstrators.

Delegation of authority within the organization must also be planned. If the chief executive officer or top official of the organization who ordinarily makes decisions is not available, a list of those next in line of succession should be known. The company official acting as the company spokesman dealing with the demonstrators will then be assured that someone is always available to make decisions.

Collection of evidence should be arranged in advance because the presentation of proper evidence to a court is necessary to obtain a court injunction or prosecute a person who has committed an illegal act. In addition, evidence is necessary to substantiate legal acts such as the issuance of warnings to insure that the rights of a person are protected.

Evidence may be presented to a court by direct testimony of a witness to an illegal act or by the use of photographs and tape recordings. The combination of an eyewitness account substantiated by a photograph of the illegal act is of most value in court.

Observers should be appointed and assigned to strategic points so they can observe and record the actions of demonstrators. Photographers, both still and movie, should be assigned to work with the observers. Sound-recording equipment should be made available so that threatening announcements made by demonstrators, warnings for dispersal, and citizen's arrests can be recorded. Each observer-photographer team should maintain a log of incidents photographed. Arrangements should be made to route all the evidence that is collected to a representative of the legal counsel. In addition, the teams should be briefed before they are posted and debriefed after they are relieved.

The contents of an announcement to be used to warn demonstrators to disperse should be prepared in writing so that it can be read to trespassers if needed. The announcement should be based on the requirements of codes or laws governing the situation. Also, if citizen's arrests are to be made, the procedure followed and the statements made to those being arrested should meet the local code requirements.

Insurance coverage should not be overlooked during the planning stage. A review of insurance protection should be made to determine the adequacy of coverage against property damage, business interruption, and liability that might result from a disturbance. For example, the basic fire policy should be examined to determine if it contains "insurrection exclusions" that might limit coverage. Also, it should be determined whether losses resulting from looting after a riot would be covered. Those questions, which are only illustrative, should be discussed with the insurance carrier. Appropriate coverage, if it is required and is available, can be secured by written endorsements to existing policies.

Planning will not insure protection, but if all routine as well as predictable decisions have been planned for in advance, all efforts during a disturbance can be concentrated on unpredictable or unusual circumstances that may develop. Also, planning to cope with disorders is a continuous process.

The plan should be tested. This can be done through simulated exercises (dry runs) during which the various aspects of the plan are checked. Deficiencies and unrealistic features of the plan can then be corrected or eliminated.

Many factors must be considered in developing a plan to cope with disorders to minimize damage and loss of assets. The items discussed are not intended to be a complete listing of all the problems that may be associated with acts of violence and lawlessness, but they should be regarded as the basis for developing an effective plan.

Terrorism

As an instrument of political action, terrorism is not a new phenomenon but has been practiced for many years—usually as a weapon of the weak against the strong to redress a balance of power. However, the incidence of terrorism activity has increased substantially and it was nearly 400 percent greater in 1979 than it had been in 1970.[3]

Terrorism has become a disturbing problem worldwide because of its wanton destruction and because so often the innocent are harmed. Also, the orderly functioning of commercial and business enterprises, institutions and governments are often seriously threatened.

A Propoganda Weapon

The spectacular nature of terrorist activities results in comprehensive news coverage and with modern communications each incident usually gets international attention. It has been said that terrorism is violence for effect. As a result, small unknown groups, through their cruel acts and indifference to innocent victims, are able to obtain overnight international attention. Although the resulting publicity may not be favorable, terrorists are not concerned with such details as long as they are able to obtain media attention.

For example, the South Moluccans were unknown until a group of them in the Netherlands hijacked a train and took a group of school children as hostages in a school house in 1976 and 1977. These incidents brought nearly all of the normal government functions to almost a complete standstill. Overnight the causes of the South Moluccans became known worldwide. Another earlier spectacular and ruthless terrorist attack that resulted in international attention was the attack on Israeli athletes at Munich. On September 5, 1972, eight Palestinian guerrillas broke into the Israeli quarters at the Olympic Games in Munich, killed two Israeli athletes and took nine other hostages. The guerrillas demanded the release of 200 Palestinians imprisoned in Israel and safe passage for themselves and the hostages to another country. After first appearing to accede to the terrorists' demands, the government then attempted to ambush them at the airport. In the subsequent shootout, five of the eight terrorists and all of the remaining nine hostages were killed. This incident shocked the world and embarrassed West Germany.

Barbaric attacks on individuals have also resulted in international publicity for terrorist groups. Two prominent individuals who were kidnapped and later killed were West German industrialist Hanns-Martin Schleyer and Aldo Moro. On September 5, 1977, Mr.

Schleyer was kidnapped in Germany by members of the Baader-Meinhof gang. He was found dead in a car trunk in France, October 19, 1977. Former Prime Minister of Italy Moro was kidnapped not far from his home in Rome on March 16, 1978. His chauffeur and four bodyguards were killed in the attack. His body was also later found in the trunk of a car. The Red Brigade, reported to consist of only 250-300 members, was the terrorist group that claimed credit for the slaying. In addition, the group claimed that between 1970 and 1978 the members were responsible for 44 assassinations, 30 kidnappings and about 2,000 acts of sabotage. The Brigade virtually paralyzed the operation of the government in Italy by seizing Mr. Moro because he had been named head of the ruling Christian Democrats and was considered to be the most important politician in Italy. This was a frightening illustration of the power that a small group of dedicated fanatics can have over an entire country.

Types of Terrorism and Terrorists

Terrorist activities fall generally into five major categories as follows: (1) bombings, (2) hostage taking and kidnapping, (3) assassinations, (4) hijackings, and (5) attacks on and seizures of facilities. Because the most frequent terrorist attack in the past has involved the use of bombs, Chapter 9 in this text is being devoted completely to that problem area.

Hostage taking and kidnapping has been a favorite terrorist technique because they can obtain money to finance their activities and they can obtain a large amount of publicity in the media over an extended period of time while demands are being negotiated. An extreme example of a hostage negotiation extending over a long period of time involved the 1979 take over of the United States embassy in Tehran. A group of students, reported to be members of the Revolutionary Front of the Islamic Republic of Tehran, took control of the embassy on November 4, 1979 along with all the embassy staff. After about 14½ months, the 52 hostages were released in January 1981. In the interim, the lives of all the hostages were constantly threatened. As a result, the terrorist were able to command world wide media attention during the entire period.

Despite the unfortunate cases of Aldo Moro and Hanns-Martin Schleyer, authorities indicate that individuals that are kidnapped or are taken as hostages have a fairly good chance for survival if they cooperate with the attackers. Of 567 individuals seized during the period 1970-1978, 35 were killed. About one fourth of the deaths occurred during the seizure attempts—usually when the victims resisted. While some victims have been held as long as six months, most have been released within a month. An unusual exception was William Niehous, an Owens Illinois Inc. executive assigned to Venezuela. He was taken from his home in Caracas by seven masked gunmen on February 24, 1976 and held until he escaped in June, 1979.

Terrorist generally endeavor to avoid the third type of activity—assassinations. Because after it becomes known that an individual has been killed, the victim is of little or no value in carrying on negotiations. Terrorists, however, will not hesitate to kill anyone in the vicinity of a victim such as bodyguards, chauffeurs, or innocent bystanders to accomplish their mission.

The fourth type of terrorist activity—aircraft hijacking— has been a very popular and dramatic technique that terrorists have utilized over the years. For example, between January 1970 and November 1977 there were 72 recorded incidents resulting in the deaths of 36 hostages and 15 hostage woundings. A total of 1,707 hostages were released and $13,580,000 was paid to meet the demands of the terrorists in 75 per cent of the incidents. The trend over the years is down, probably because of an increase in security controls.

With regard to the last category of terrorist activity—attack and seizure of facilities—there were 290 recorded incidents of this type during the period 1970-1977. A total of 358 were killed and 332 wounded during those attacks. Facilities attacked were: government offices, banks, military installations, and residences of individuals.

Authorities in this field generally group terrorists into five categories. These categories are determined by the motivation of the individuals or groups participating in such activity. They are: (1) political, (2) minorities, (3) criminal, (4) mentally disturbed, and (5) religion. It is not always possible to easily determine in which category an individual or group should be placed. In fact, it is conceivable that an individual could be placed in all five.

Political terrorist groups such as the Black September, the Palestine Liberation Organization, the Puerto Rican FALN and the Red Brigade attempt to influence government activities and gain power through extortion and blackmail. Negotiations with this category is generally difficult because this type is most likely to select victims or incidents to generate international attention and to make demands that are completely unreasonable and impossible to meet. For instance, terrorists groups in the past have demanded that those belonging to their organizations being held in jail be released and that large sums of money be paid. In the case of Aldo Moro, not only the release of prisoners was demanded but $1 billion in ransom was also specified.

The Black Liberation Army in the United States and the South Moluccans, previously mentioned, are examples of minority groups that have attempted to gain recognition and motivate change by engaging in terrorist activities. Such groups usually feel they are being discriminated against because of economic status, religion, race, national origin, etc.

The criminal terrorist is usually an opportunist who recognizes that by making an attack on an individual or individuals it is possible to obtain money. Such an attack may take the form of kidnapping and a demand for ransom or it might be a threat to do bodily harm unless money is paid. The activities of political terrorist have suggested to criminals that this is an easy method to use to obtain a large sum of money.

Mentally disturbed individuals are those that attack others to satisfy an imagined wrong or an obsession while the terrorist motivated by religious zeal is typified by the Black Muslim sect that seized government buildings in Washington, D.C. in March, 1977. Twelve members of this group seized 134 hostages in three buildings and, while constantly threatening the hostages with death, held them for nearly 40 hours. One hostage was killed and a dozen wounded before they were released the morning of March 11, 1977.

Prevention

Preventing attacks by terrorists or making people or facilities less attractive as targets is regarded as the best defense. Such attacks can often be prevented by developing defensive plans to counteract their efforts. A terrorist is typically characterized as one who has little patience, likes to act quickly in a violent way and does not like to wait until conditions are right for an attack. As a result, a carefully developed plan to include a series of obvious protective safeguards or counter measures will often discourage an attack so that another target with less protection will be attacked. Of course, if the terrorist is intent on attacking a particular target because it will be the only one that will allow the objective of the group or individual to be realized, then the attack may be carefully prepared and executed in spite of the defensive measures in place. The attack on Aldo Moro, previously mentioned, is the type of attack that might have been made regardless of the measures taken to protect him because he was the key to the objective of the terrorists. However, protective measures taken to protect Mr. Moro were not adequate. The terrorists knew this and were able to make a successful attack taking advantage of the weakness in the security measures while he was enroute to his work place. Any protective measure adopted must therefore be effective so that even if an attack is made, actions can be taken to protect the target.

It can be anticipated that whenever possible, terrorists will make a surprise attack. As a result, a victim must be prepared at anytime to respond to such an attack with defensive actions. An attacker normally will not expect defensive actions or the utilization of countermeasures when making an attack. As a result, such actions may surprise the attacker to such an extent that the attack is broken off.

Terrorist attacks can usually be expected to take place at a work place or other facility, at a residence or while a terrorist target is in transit. In a short chapter such as this it is not possible to go into any great detail with regard to protective measures that should be considered for proper protection of terrorist targets. As a result, only the vulnerable areas to be protected are highlighted. Other reference works that detail protective measures should be utilized to develop a plan to neutralize terrorist attacks.

Adequate planning should include, as with other protective measures adopted in any organization, the endorsement of the plan by the top policy making executive or group. A policy should be adopted and fully supported by the operating executive or group.

REFERENCES

1. *Disorders and Terrorism,* Report of the Task Force on Disasters and Terrorism, National Advisory Committee on Criminal Justice Standards and Goals, Washington, D.C. December 1976. Pages 1 and 15.

2. Report of the Committee on Government Affairs, United States Senate. Report No. 95-908 to accompany S. 2236, An Act to Combat International Terrorism. Washington, D.C. May 23, 1978.

3. *International Terrorism and Business Security,* The Conference Board, 845 Third Avenue, New York, N.Y. 10022. October 1979. Page 3.

Self Test Questions

1. What is the difference between a crowd and a mob?

2. What is the definition of a mob?

3. Name the typical actions that can be expected from a mob?

4. What are the psychological factors that may influence mob behavior?

5. What are the four phases of a civil disorder?

6. Should the top official of an organization meet with demonstrators?

7. How is terrorism used as a propaganda weapon?

8. What are the five major categories of terrorist activities?

9. What is the best defense against terrorist attacks?

= 6 =

Computer Security

AMERICAN industrial and business organizations have been rushing headlong into the use of computers. Expenditures for 1980 were estimated at $71 Billion.[1] However, there is little evidence that enough serious thought has been given to the protection of computers and the information processed by them. In spite of the marvelous feats it can perform, a computer may actually cause serious losses if proper security is not planned.

The first factor that should be clearly understood is that the computer is a machine and not an intellect. It has no conscience and no moral or ethical code. It will do precisely what it is directed to do, and in the hands of a dishonest but competent individual it has an enormous capability for economic disaster. That and other hazards to be discussed later could destroy an enterprise.

Another important factor is "the computer mystique." Because the operation of the computer is seemingly complicated and difficult to understand, many executives have developed a false feeling of security. The computer is a mystery to them, and so they assume that very few people have the capability of understanding how it operates. They overlook the fact that there are a number of classes of employees who understand the operation very well—programmers, operators, and analysts, among others. Those employees are not overwhelmed by the computer mystique and so are in an excellent position to take advantage of a situation in which false feelings of security have deferred the institution of a computer security program.

The basic step in designing any computer security program is to make an analysis of hazards or risks. In making the analysis, each organization should initially determine what information is handled by its computer and how it is being processed. Hazards to the computer and its information will exist on three levels, and each level will have risk aspects that relate to the data themselves and risk aspects that relate to processing.

Security vulnerabilities that affect computers can be divided, most basically, into three classes: (1) those that threaten the physical integrity of the computer installation and its contained information, (2) those that threaten the loss or compromise of the data from outside the computer site, and (3) those that threaten loss or compromise of data from inside the computer site. The three levels of risks will be separately considered.

Physical Vulnerabilities of the Computer Center

The chief risks here are five: (1) fire, (2) acts of sabotage, (3) industrial accident, (4) natural disaster, and (5) mechanical or electrical malfunction of the system.

Fire

The most serious and most common physical danger is that of fire. All the most significant computer equipment including the main frame and principal peripheral items like disk drives, tape drives, multiplexers, and various input-output devices are electrically energized. There is the constant possibility of circuit failure, electrical overcurrent conditions, insulation fires, and related combustion problems connected with electric energy. Because of the extensive damage that prolonged heat and uncontrolled combustion products such as smoke, particulate matter, and chemical changes can do to sensitive electronic components, most attention to date has been paid to this aspect of security. However, despite the development of fire protection doctrine and the availability of a variety of automatic detection and extinguishment systems, many large computer installations are relatively unprotected, even this late, against fire loss.

The National Fire Protection Association (NFPA) publishes, among its 10 annual volumes of National Fire Codes, Standard 75 for the protection of electronic computer/data processing equipment. It should be the basic guide to fire prevention and extinguishment for all computer installations. The NFPA is internationally well known, and it has led in fire safety standards development for over seventy years. Standard 75 is technically oriented but easily understood and should be read by any manager with responsibility for the computer. In barest outline, the standard requires that computers be separated from other occupancies by fire-resistive barriers, that adequate extinguishment equipment, preferably automatic, be provided, that all assigned personnel be trained in the use of the equipment and in all emergency response actions in the event of fire, that building and design features be adopted to minimize the communication of any fire that may start, and that certain housekeeping and administrative practices be instituted to reduce the causes of fire and potential losses if fire should occur.

Sabotage

Computer sabotage is a real and present danger, not just a dramatic possiblility. Early samples of sabotage damage of $1.6 million at Sir George Williams University in Canada in 1969 and of $6 million at the University of Wisconsin in August 1970 focused on the computer center. Extensive sabotage of computer operations has also occurred in the past several years at the Dow Chemical Company in Michigan, at Boston University, at Fresno State College in California, and at the University of Kansas. In the last case, elementary security planning greatly reduced the damage that might otherwise have resulted from a bomb that exploded in a stairwell near the computer center and injured three persons. Because the computer is inherently so vulnerable and because it seems to be the key symbol of the "establishment" to many ideologically militant dissidents, it will continue to be the target for sabotage.

Sabotage can be as obvious as a bomb or as indistinguishable as the application of a small electro- or pocket ferromagnet to magnetic storage media. Blows with a hammer or other heavy instrument on components, use of caustic solutions on circuit boards and wires, and introduction of foreign matter into mechanically moving parts can all do a highly effective job of destruction. Even more subtle is the deliberate but limited garbling or manipulation of information to render an entire computer operation unreliable because of random errors. That technique can have long-lasting effects because computer operations can be the basis of ongoing activities elsewhere before the basic problem is discovered. Entire projects such as batch analysis in quality control or sales order administration reports can be ruined in this way.

The basic vulnerability underlying sabotage is that of unauthorized access to the computer installation. An effective system of access control in which positive personnel identification coupled with "need to go" and tight supervision of computer operations in the main-frame area will reduce this vulnerability to a low level. The other precaution for its control is adequate personnel selection and screening to assure a high level of confidence in the trustworthiness and reliability of those who are permitted regular computer access. Even with adequate hiring screening, however, there will be other vulnerabilities connected with access to the computer area, and further admittance controls are required. They are discussed in the later sections that deal with internal threats.

Industrial Accident

Aside from fire, there may be major industrial accidents that will affect the computer. Selection of a computer location that is not contiguous to a high-hazard area is therefore an important preliminary consideration. Accidents that have led to considerable computer damage include building collapse, explosions, and interruptions in utility services such as electric power and air conditioning. The electrical interruption can be particularly serious in systems that are operating in a real-time mode and for which constant power is a requirement. Even momentary surges (swings above or below the rated current) can do serious damage to programs that require constant power because the internal behavior of processing circuits will be changed. Loss of air conditioning can result in more than discomfort. The distortion temperature of magnetic tape is around 140°F. At that temperature random changes in the magnetic character can occur and produce information changes. Sustained temperatures in excess of 140°F can also cause random malfunction of internal parts of the central processing unit and other energized equipment. Continued operation without controlled air exchange will rapidly increase temperatures until they are at or beyond the critical temperature in certain locations.

The principal solution to utility failure problems is provision for an independent, self-contained emergency power generating capability. The power generated should be of sufficient rating to provide the minimum needed to maintain any level of operation likely to be in process and also provide a margin of safety. Whether constant power or merely standby power is required is a determination to be made by the data processing manager and the building services manager or plant engineer. In the determination of the standby power and its switch-in, careful consideration should be given to the maximum allowable period of no power and to the availability of competent technical personnel to troubleshoot problems during such a period. If there is no need for constant power, then the tolerable down time may be anything from a few moments to a few hours. If competent technical personnel are available to diagnose the trouble immediately, it may be preferable to rely on manual rather than automatic switchover to standby power. That will prevent the unnecessary switching for readily correctable difficulties. If there is any question of the availability of technical personnel, or if the down time limit is very short, automatic switching may be required.

Natural Disaster

Floods and earthquakes are among the catastrophes most likely to have a serious effect on computer installations. The availability of real estate or its cost of acquisition is not the only factor to consider in locating a major computer operation. The whole environmental context is important. To the greatest extent possible, installations should be in areas without histories of or predispositions to natural calamities. Low-lying, river basin land along regularly flooding rivers would be a poor choice. Coastal property within the hurricane belt would be another poor choice.

Often, the usual property site selection factors are applied to computer installation sites to determine feasibility. Among those factors are accessibility by public transport, availability of public utilities, labor market, and make-ready or development expense. Some of those factors may be less meaningful for computer installations when the entire store of information belonging to the enterprise will be exposed through the computer. Electric power generation on a purely local and proprietary basis, although more expensive than purchase from a public utility, may be justified by the additional reliability and insulation from power system difficulties. Some remoteness, particularly from major metropolitan population concentrations, may be highly desirable from a physical access point of view. Remoteness such as to be out of the response range of police and fire departments can, however, be even more serious than being in too dense an area. Adequacy and availability of emergency services must be evaluated in the plan. The use of automatic fire and disaster control resources and of integrated security systems of the kind discussed in Chapter 10 can do a great deal to lessen dependence upon municipal emergency service facilities. Inclusion of those systems at the time of initial construction is far better than adding them later because it is cheaper and because the system can be planned for optimum performance if it is designed at the same time as the computer center it is to serve.

Both disaster planning and fire protection dictate a reliable
system of multiple-generation information retention. Two or
three generations of data, with a capability of reconstructing
the changes made from one generation to the next, will permit
reassembly of the data base, programs in process, and program
instructions if a disaster should damage or erase the information
then in the computer. But multiple-generation data maintenance
requires that the link between old and newer data actually be
kept. Old data without a basis for updating them to match what
is now in process are much less valuable and may not prove
to be the safeguard they are expected to be if an emergency
requires rapid reconstruction. In addition, the storage of the
backup data should be under circumstances not affected by common disaster threats to the primary information. That means
separate storage sites with secure storage of each set of data
at each site. It also means a flow of updated information to replace each generation of backup data on some periodic basis.

Perhaps the most serious problem in terms of potential disaster lies in finding temporary alternate equipment. There is a
popular belief that if you know where another facility has the
same basic main-frame equipment, you can rely on that facility
as an emergency backup to your own. That is not so for at least
five reasons, each of which must be considered when alternative
emergency facilities are planned. First, the machine configuration
at the other site may not be compatible with yours. For example,
if the memory core is smaller or if there are fewer printers, your
operations will have to be modified before it can be run. Second,
there may not be any open time on the backup facility when
you require it. The objective of having high-cost computers is
to use them as much as possible to increase the pay-back. All
users are striving to reach that objective, and the passage of time
generally means that there will be less and less open time on
any given computer. Third, there may be a software compatabil-
ity problem when a company tries to run its programs on the
alternate hardware. For example, the operating system (OS) for
the two computers may be quite different. Fourth, the information flow to the computer may depend upon telecommunications

resources that are not available at the alternate site or that are devoted to the needs of the alternate site proprietor. Fifth, if loss of your computer has resulted from a common problem such as failure of the public utility or natural disaster, the alternate facility may be in the same position. As is true of alternate storage sites for recorded data, alternate sites for emergency computer backup should be out of the radius of common disaster.

The most effective planning technique for assuring that backup needs have been recognized is to develop a matrix of all the programs being run or planned. The matrix will rank the programs in order of criticality to the particular enterprise. For each program will be shown the machine configuration, the scheduled machine time, the processing routines, and any special factors that make the program difficult or impossible to duplicate. Then will be shown the locations (if any) of backup resources, the most recent date on which the backup was contacted to confirm availability, the switchover time required to get the program on the alternate hardware, and the limits on the permitted usage. Such a matrix, updated regularly, will identify the critical programs and will indicate whether there is any backup at all available for them. The exercise may serve only to establish that there isn't any practical emergency backup. In that case, the need for doing a thorough job of on-site security planning will be obvious. In either event, the matrix planning should be done to give management the required decision options.

There are recorded cases of things like eccentric tape drives, bent capstans, scratched surfaces, and broken tapes that have caused serious dollar loss. The smaller the malfunction is or the more difficult it is to observe, the greater the ultimate problem because of the vast amount of data that may be processed under faulty conditions.

In addition to mechanical and electrical malfunctions, preventive planning must consider human malfunctions. Operator error, for example, can result in the erasure of vital data that can't be reconstructed. Aside from assuring complete familiarity with all operating routines and security regulations on the part of all personnel assigned to any duty in the computer operations area, it is most important to preserve absolute accountability in terms of machine and data access. That means that there should be an unbroken chain in which the identity of each person having had access to either information or machines is shown, together with the date and the time in, time out. For the main-frame area that can be tied to the machine process sheets and metering records so that all machine operation is identified by program run, personnel involved, and time required.

Maintenance of standards for program running time and the insistence of complete "exceptions logbooks" will permit later analysis and comparison to be certain that programs that have been run have not disguised the use of the machine for other operations as well. The problem has been a real one and has resulted in unauthorized use of computers to do unrelated work (as when the operators were contracting with outsiders to run their programs for a consideration) and when unauthorized operations were conducted to manipulate data from other programs within the enterprise in the course of theft or fraud.

Threats to the Data from Outside the Computer Site

The second major category of security vulnerabilities can involve direct telecommunications links between the outside locations and the computer, as in time-sharing operations, or it can involve loss or compromise of data by surreptitious means. The latter can be discussed first because it is probably the lesser problem and the one capable of the most-straightforward solutions.

Surreptitious Data Theft

Theft of data can be motivated by desires to learn trade secrets, to anticipate a competitor's market or financial strategy, or for any of the reasons that would lead to conventional industrial espionage. The difference between vulnerability through the computer and vulnerability elsewhere is that in enterprises that have computerized most of their operations, the computer, at some time, will probably be processing the target information. If undetected computer data access can be achieved, then a knowledge of computer schedules will give the attacker all the planning information he requires.

One technique for data theft is the capturing of electromagnetic radiations from operating computers and the analysis of the signals for retrieval of data. This is simply a form of intercepting an ongoing communication by tuning in on the signal frequency used to move the data either from unit to unit within the computer center or by data link to some other location. It requires that the interceptor get close enough to the center or the communication link, that he have intercept gear that can monitor the spectrum and identify the particular frequency being used for his target information, and that he have some way to demodulate or convert the electronic signal into plain text. He need not do everything from his access vantage point. If he can record the target transmission, he can carry the recording away and reduce the information to plain text at his leisure somewhere else.

One defense to the technique described is the encasing of the computer operation in protective electronic shielding that will suppress the radiation of energy beyond the physical screen. Because it is expensive and requires major physical installation activity, electronic screening has been reserved to date for the most sensitive computers running highly classified defense information. However, when a proprietary risk is sufficiently critical, shielding should be considered, particularly as part of a construction plan for new computer centers. Shield design and installation technique are of central importance, and there is a low error tolerance. A very expensive shielding job could be rendered useless through small design or installation errors that permitted some leakage of target frequencies.

Another technique, and one that is particularly relevant when communications links are the targets, is the hardening of the link itself or the processing of the transmitted data to make them difficult to capture and recover. Link protection involves shielded cables when hard-wire connections are in use and random frequency changes when microwave or radio wave is used. Processing the data themselves to make capture difficult involves two techniques, one more costly and complex than the other.

The less complex technique requires that multiple communications be transmitted simultaneously over the same link. To recover any single communication, the receiver or interceptor must be able to filter all the other signals out and isolate the one he wants. That may be so difficult or expensive as to preclude any serious effort to do it. A modest precaution that features the simultaneous transmission of nonsense signals or even mixed legitimate traffic over the target link can neutralize most interception of data outside the computer center. The other technique involves the design and installation of crypto systems for encrypting and decrypting the data. Crypto systems are expensive and can increase the possibility of error because of the frequent changes of the form of processed data. They were formerly used only in computer systems that processed political and military data of intelligence value. However, crypto systems are now available for business use, and they should be included as planning items for computer installations.

One other point is important in the consideration of the vulnerability of computer operations data to outside interception. The closer the attack point to the initiating or receiving terminal, the easier the job. Conversely, protection is needed more at the points of origin and receipt of computer data being transmitted to other locations than anywhere else on the transmission system. If the transmission is by switched network telephone lines, then the multiple signal transmission spoken of earlier will automatically become a protective feature when the computer signal joins others on the common carrier lines. Moreover, choice of specific sets of wires for transmission of particular signals over the telephone company networks is dependent upon system load. Automatic switchgear will select the optimum transmission path at the moment in time when the message must be sent. There is no way to know in advance what path that will be, so an attacker has the dual problems of multiple signals and unknown wire path. That makes any serious vulnerability very unlikely.

But, if the attacker can gain access to the wires inside the computer or the receiving terminal, he can set up a tap and retrieve the data. That is no different from any other telephone tap and involves either physical or inductive connection between the target wire and the surreptitious listening gear. The connection is possible at any point along the target wire where it is accessible and identifiable. The leads from the computer center to the various telephone terminal boards and from the terminal boards into cables going out of the facility are all sensitive. Adequate physical security must be maintained over the wire from junction points outside the facility back to the computer on one end and to the input-output terminal on the other. For the most part that will involve locking terminal or frame rooms, enforcing strict movement controls on persons entering areas that contain communications switchgear inside the facility, and cooperative arrangements with the telephone company to provide acceptable protection to some junction points on the premises perimeter. The protection applied will be some combination of lock control and intrusion alarms. Underground service connections are much safer than overhead wire connections.

Time-sharing System Dangers

Of all the dangers to data compromise from outside the computer area, those presented by time-sharing systems are the most serious. They are serious because error or system malfunction as well as intentional espionage efforts can disclose data and because more and more time sharing is being planned.

Basically, a time-sharing system is one in which several users have access to a single computer in which one or more programs are being processed. The multiple access may be from within a single enterprise, as when various departments have access to a central computer, or it may involve unrelated users, as when a service bureau or central commercial processing facility uses a large computer to process several programs for different customers. The ultimate combination is the multiprogram, multiprocess time-sharing operation in which many unrelated users have access to a single computer simultaneously processing many unrelated programs. That situation is now a common one.

The basic problems are three: (1) assuring that only authorized persons can key into or have access to a program, (2) preventing the occurrence of a system malfunction from accidentally disclosing data, and (3) being certain that authorized access does not result in unauthorized change or manipulation of data.

The most common technique for limiting access to authorized persons is the use of a code or lock word; the person attempting to gain access to the computer must identify himself by a code word or symbol. It can be as elaborate as needs require, including a separate code word known only to the person who will enter it together with some other item of personal identification such as name or payroll number. The computer will look up the two items against a stored memory table to be sure they are the two items required. By making the actual code word or symbol a truly random one not related in any way to the personal identification data, a high level of reliability that the person gaining access to the computer is actually the authorized person is achieved.

The precaution is subject to intentional defeat if the authorized person deliberately or negligently discloses his code word or if persons at the computer end disclose the code word to confederates outside. The second possibility is more related to internal compromise threats, however, and it can be dealt with by techniques recommended for that problem. Another possibility of compromise of code words is failure to program the computer to suppress them and not repeat or print them out for recovery elsewhere in the network.

A further precaution against unauthorized access is the requirement that both a physical device and a password be used. For example, a card could be coded with information in magnetic format to be inserted in a card reader that would control a logic interlock either with the electrical supply or an intrusion alarm system. Before the computer terminal could be energized, first the card device and then the individual code word would be required to gain access to the computer. For remote input-output stations such as offices in rented space in foreign cities, the technique may be highly useful. With intrusion alarms at either the premises or the computer input-output terminal itself, a fairly secure condition against unauthorized computer access can be established.

Finally, access from remote terminals can be managed to limit individuals to preassigned programs. Their personal code words will authorize only certain program access. By developing the monitor or supervisory program at the computer itself to signal any attempt to gain access to a program not authorized for that user, to stop all data output to the terminal, and to print out the attempted use transaction, an immediate response to unauthorized data access can be made. Each extra security step in this chain will add to overall cost, mostly software cost because of the additional programming required. The extent to which the routines are carried must be determined on the basis of criticality and cost trade-off—the allowable cost of any countermeasure considered against the size of the possible loss and the availability of alternate countermeasures.

Loss Threats from Inside the Computer Installation

The greatest vulnerability to losses, both of information and of assets through manipulation of information, is in the computer center itself. Because of the large number of persons who normally have access requirements of some kind and the vast amount of in-transit data to and from the computer and various users, data are always exposed. They and the computer facilities processing them can be used to perpetrate frauds, commit thefts, and engage in industrial espionage.

Fraud and Embezzlement

A variety of computer embezzlements have already been recorded. among the more dramatic was the diversion by Stanley Mark Rifkin of $10.2 Million from the Security Pacific National Bank in Los Angeles in October 1978.

A more typical case is illustrated in the 1979 inventory shortage concealment suffered by E.R. Squibb and Sons in the amount of $1 Million at the hands of a Squibb plant computer manager.

There was also the case of the back office investment house employee who embezzled over $250,000 over a seven year period, becoming a Vice President before the thefts were finally discovered.

Many methods can be employed to program a computer to defraud. A single change in a computer program can convert abnormally high inventory losses to breakage. Material reported as lost can then be removed without the theft being noted. The evidence can be destroyed by simply reversing the entry. Or a data processing employee in a financial institution might round off fractions of cents in interest calculations and transfer the amount collected to his own account.

A computer payroll system is also a source of fraud. Fictitious paychecks can be printed or extra wages and overtime can be programmed to be paid to designated individuals. Or a few cents of income tax deduction from each pay check in a facility can be programmed for temporary collection in an account that would be set to pay out to a payee in collusion with the fraud.

Theft

Computer installations are sometimes an invitation to steal. Because there are no security controls, a thief can quickly and easily remove or copy valuable records or tapes that he can sell later. A tape can be copied in a few minutes. It could contain a secret customer list, priceless product formulations, or sets of valuable operating procedures. Payroll information, inventory records, accounts receivable, accounts payable, or other ledger data might cost a company thousands of dollars if stolen.

Theft of another kind must also be considered—theft of computer time by employees. That can occur when the actual use made of a computer is not supervised or audited. There have been cases in which computer personnel were running entire sets of business records for themselves or friends.

Espionage

Because computer systems are now utilized to store information critical to the operation of the company, espionage is a prime hazard and can most easily be committed from within the installation housing the computer. Copies of tapes or of printed output reports may contain concentrations of data that would otherwise require weeks or months of effort to accumulate. In industries with proprietary processes, the process control data may be available from process control programs or from computer-generated management reports. The person bent on espionage can acquire his target information either as hard-copy printed material or in machine-readable format, as on a magnetic tape or disk. Access either to unprotected information output or to an unsupervised computer facility will be enough for the skilled spy. When stolen data have been taken in such a way as not to leave any evidence, the loss is doubly expensive because the victim enterprise continues to operate on the assumption that data are safe. If the compromise were known, additional expenses might be avoided even if the data could not be retrieved.

Computer Installation Safeguards

Access Control

A physical control system should be designated to prohibit unauthorized personnel and visitors from entering the computer area. Some companies view their computer installations as showplaces. They encourage visitors and often fail to provide minimum security precautions. Such companies have not considered the possibility of the hazards already discussed and the potential serious losses that might result.

Controls on the computer center should deal separately with two control periods. One will be while the center is in use; another and different type of control will be necessary when it is not in use. During the period when the center is not in use, the entrance and all other openings should be securely locked. Also, the walls, ceiling and the floor must be so constructed that surreptitious entry is not possible and forced penetration will be difficult and obvious. An alarm system to signal area intrusions will give added protection. Periodic inspections should also be included in the security plan.

When the computer center is being used, it is the usual practice to secure the entrance with a lock and to designate someone to control those entering and leaving. It is possible to control entrance in a variety of ways. If the center is occupied by only a few employees, a supervisor on site might be assigned the responsibility for entrance control. A telephone or other audio connection might be installed outside the center. Or a doorbell might be installed to signal the center that an individual on the outside wants to communicate with someone inside.

A more elaborate security control would include closed-circuit television and a communication circuit between a camera outside and a monitor inside. A remote electric lock release or digital lock might also be added to permit self-admittance or remote activation.

If the area is too large or the traffic through the entrance too heavy to be controlled from inside the center, a working clerk or typist might be placed at a desk outside the entrance and be assigned the duty of controlling entry. Another application of closed-circuit television would tie in cameras as well as other controls at the entrance to a security control center. The entry would then be managed by the security organization from the control center.

Regardless of how the area is controlled, an access list indicating those authorized to enter is essential. The list should be in possession of the employee controlling the area and should be accurate and current. Fully automated access control is possible through the use of card readers, door interlocks and stored memory authorization lists.

Internal Security Controls

Simply because an individual works in a computer center, he should not have free access to all its areas. Programmers, for example, usually will not require access to the computer controls. On the other hand, the operators do not usually need access to areas where files are maintained. Areas within the computer center should be so designed that there is adequate control on those who work there.

Because program and data files are critical items in any computer system, a storage library should be established and procedures should be implemented to give protection to the storage area. Files should be removed from the library only when they are needed. Only authorized personnel should have access to the library. Records of file use should be kept, and a file should be checked in and out of the library area by name of the person taking the file.

Employees within a computer center should be rotated. For example, a programmer who has been responsible for a sensitive program for a period of time should be shifted from that program and assigned to another. Employees should be aware that they will be transferred in this way so that they will not be motivated to tamper with programs and manipulate them for their own benefit.

Only one person or operating group should be responsible for an operation at any one time. Ideally, that means drawing lines between the employees who authorize a transaction and produce the input, those who process the data, and those who use the output for reports or for other management purposes. The same controls should cover scheduling, manual and machine operations, maintenance of programs, and related functions. Programmers should not have access to the entire library of programs, if only to guard against the possibility of malicious damage. Also, programmers should not operate the machines. If duties are properly separated, the possibility of damage is minimized. Each employee will have only a limited role in the entire system's operation.

It is essential that written instructions for all personnel in the computer center be developed. The instructions should be kept complete, current, and understandable. Supervisors and security personnel should constantly check to insure that the operations personnel are following the procedures outlined.

Internal Program Controls

Despite all efforts at physical protection of the computer center and its environment, there will be errors or deliberate attempts at information manipulation. The potential harm from both can be sharply reduced by controls built directly into the programs that will signal an error or unauthorized exception. Specific development of some of the controls discussed here is a programming task, but assuring that the controls are considered and used is a management task. The work of a programmer in developing the form of a specific control can readily be checked by another programmer or more senior person in the systems organization to assure it is properly done. Control can be achieved as follows:

Require changes to master memory to be made by different personnel than those who handle day-to-day operations to reduce the opportunity for fraud. If the person who handles the daily operations cannot change the master file and the person who changes the file does not have regular access to running programs, there is less chance that either could make a change in the data to his own advantage.

Document master data change inputs by requiring authorized signatures, serially numbered and controlled forms, and retention of the authorization document until later verification of the changed program.

Use limits checks to assure that a given transaction cannot occur if its base or computed data exceed stated numerical limits such as "no interest payment above $1,000 or no paycheck over $2,000."

Set up bounds registers to assure that access to areas of the core or auxiliary memory cannot be gained during programs that do not require those areas.

Check out new programs before allowing them to process current base data.

Use batch and hash totals to assure that all required transactions have been performed and unauthorized transactions have not. The totals will utilize input data to verify the process. For example, if a given number of sales orders have to be processed, the numbers of the individual orders can be totaled and the total can be used to check the program output. If all the orders were handled, then the order number of each will be retained for the addition. An error in the comparison of after-process order number batch total with reference order number batch total would indicate an omitted or inserted order. Another illustration would be the addition, by category, of various components in a program. In a payroll program, for example, state withholding, federal withholding, social security, and other tax deductions should total to a determined number in a correct program run.

Use time and error logs to record the actual time used on the computer. Program standards, mentioned earlier, will indicate how much time is required for a normal run. The machine log, showing opening and closing meter numbers, will indicate how much time was actually used. If it exceeds the standard time, then the error log should contain an explanation of the additional time used to correct an error. This technique minimizes unauthorized machine use.

Insurance

Constantly increasing investments in program development, computer hardware, and stored data indicate that every organization should evaluate insurance coverage. Damage to a computer and related equipment and files is like an iceberg; the part that can be seen may represent hundreds of thousands of dollars, but there may be other, even larger losses to be considered.

After all the steps already outlined to protect equipment and the information system have been taken, insurance must still be considered as catastrophe backup protection. There are four possible areas of financial loss to consider in buying insurance: (1) loss or damage of equipment, (2) cost of reconstructing files and programs, (3) other costs incurred in returning to normal operation, and (4) any business losses incurred by disruption of normal business. The insurance industry has recognized the needs of electronic data processing and will provide coverage to organizations will institute acceptable protection and backup procedures. In determining insurance requirements, the following items should be reviewed:

- Is all equipment completely covered for any loss?

- Does insurance coverage include the loss of recorded data as well as the cost of new hardware? Will it cover rebuilding data?
- Is business-continuation coverage included to pay for temporary operations at another location?
- Does business-interruption insurance protect against forced shutdown because the data processing department is disabled or destroyed?

REFERENCE

1. *Computerworld,* Dec. 19, 1977, pg. 49.

Self Test Questions

1. List some key vulnerabilities of any computer center.

2. Which NFPA Standard among the National Fire Codes deals with computers and data processing?

3. Why is attention to electrical power supply important?

4. Name some techniques or methods of data theft from EDP installations.

5. What is the basic element of physical protection for the data center?

7

Prevention of
Frauds and Theft

IN dealing with the subject of prevention of fraud and theft, the first concern is the probable size and economic impact of the problem. Many business managers refuse to believe that their enterprises are victims of criminal fraud and theft. However, even cursory reading of the most recent data should alert every executive to the critical need for serious attention. Some of the more impressive of the findings include the following:[1]

- Annual losses from *non-violent* business crime may exceed $33 Billion.[1]
- Unreported commercial theft losses are more than double those of all reported private and commercial thefts.[2]
- Employee theft, embezzlement, and other forms of business crime exceed $2 billion annually in the United States.[2]
- Over one-quarter of firms employing more than 1,000 persons acknowledged in a survey several years ago that employee theft of materials, products, and tools was a serious problem.[2]

- The smaller the business, the greater the impact of loss. Businesses with receipts under $100,000 lose 3.2 times the average national and 35 times the average for businesses with receipts over $5 million.[3]

Theft losses are on an uptrend.[6] All reported crime shows an increase annually. Unreported crime will increase at least as often as reported crime. With each passing day, the business loss picture from frauds and theft grows worse. A responsible attitude about loss control prompts the manager to ask four prime questions:

1. What forms can the losses take?
2. What are the causes of fraud and theft losses?
3. Where will the losses occur?
4. What can be done to prevent and detect losses?

The Form That Losses Take

Theft losses can be money, property, sales and markets, personnel, reputation, and goodwill. In some situations a single loss can involve all elements.

The most common losses are of money and property. Other losses (for example, the loss of a market through the effects of successful industrial espionage) can be more extensive. Losses of reputation and goodwill often follow upon the inability to deliver or render timely performance for customers. An illustration of that type of loss can be seen in the international air transport industry. Particular carriers, because of repeated theft losses of high-fashion merchandise such as imported shoes and dresses and of high-value jewelry and watches, found that regular shippers were shifting business to other carriers. Losses of personnel involve not only the key man pirated by another firm but the apprehended thief who must be discharged. All the costs of processing, training, and supporting him and his replacement should be considered part of the cost of theft loss. Applying the tests on criticality, suggested in Chapter 1, to theft losses will soon indicate how much more there is to what at first appears to be a straightforward property or money theft.

Even the most commonplace items can be subject to theft and add to the loss total. Among unlikely thefts reported are $15,000 in brooms over a 6-month period, $50,000 in cardboard boxes over 3 years, and $50,000 in non-narcotic pills from a pharmaceutical house. Frequently the opportunity to steal an item will result in its theft even though the thief has no plan to consume the item himself or sell it to another. A classic case of theft of that character involved an estimated $5,000 worth of various spare parts, bits, pieces, and items with virtually no usefulness or value apart from the work project from which they were stolen. They had been squirreled away over a period of years by an employee who simply collected the stolen material in his cellar where a good deal of it was recovered after his discovery.[4]

Some items are stolen in cycles or seasons. It is common experience among security personnel to find the theft of pencils, pads, and sundry office supplies increasing in September—coincidentally with the opening of school. It is also usual for retail stores to find larger theft losses from shoplifting on late store nights, during the pre-Christmas buying rush, and after 3:00 P.M. during the school year. The last finding can be correlated to the increasing incidence of shoplifting among teenagers, many of whom are still in school and, until caught stealing, ostensibly law-abiding.

While high-value, small-size items will constitute target theft risks (for example, jewelry, watches and electronic components) almost anything that is portable will be someone's theft target. Opportunity is the prime factor in theft. Motive is the other. But it is almost impossible to divine the variety of motives different persons can have for stealing. Often they are deeply interwoven with personality difficulties and emotional problems and bear little or no relationship to profit. With so many unpredictables attaching to motive, the factor most amenable to controls is that of opportunity. It is here that the most leverage will be available for the application of countermeasures and that the most attention should be directed. Specific techniques will be suggested later for the reduction of theft and fraud opportunities.

Causes of Fraud and Theft Losses

The most common causes of fraud and theft are negligence and naiveté—negligence through failure of enterprise management to examine theft vulnerabilities critically or to take preventive action when indicated, naiveté because of some common assumptions about theft and honesty, almost all of which are wrong.

Naïveté

The first wrong assumption is that production or hourly paid workers are more likely to steal than others or that managers and executives can be relied upon not to steal in most cases. In fact, more serious thefts are attributable to managers and executives than to any other class of employee. Embezzlement, a crime involving only persons in positions of trust, accounts for $200 million in estimated annual losses. Earlier studies show that, in the limited area of embezzlement from banks, presidents, vice-presidents, managers, and cashiers were frequently the culprits.[5]

Thefts are more likely to be committed by persons who have discretionary authority and are in positions of trust than by others. First, such persons are familiar with protective routines; they know what is likely to be discovered and what is not. Second, they have regular and continuing access and opportunity. Unlike wage workers, who are often subject to question when they are away from an assigned work area or are involved in any but an assigned task, supervisors and executives can be many places and involved in many activities. Third, the opportunities to get into difficult personal financial positions are often greater. A company officer can usually borrow more than his subordinates even though he may not be in any better position to repay when the loan is made or falls due. Fourth is the belief, frequently well founded, that a firm will attempt to conceal rather than expose a theft involving a senior employee. The fear of punishment is often not really a deterrent. Finally, there are the ideas that there is something inherently more honorable about a supervisor or executive than about a line worker or that a hard worker is necessarily an honest one.

Although well entrenched in business management circles, all those ideas are really just reflections of a middle class self-image. The only reliable assumption about who will steal is that anyone may, given the opportunity. There is no predictive gauge for selecting the thief in advance.

Negligence

The negligence factor, the most important single cause of theft loss, can run from such routine oversights as failure to lock a drawer or door at quitting time to more bizarre bloopers. Permitting plant entry to a person whose pass contained a picture of a gorilla or assuming a visiting industrial spy was a building inspector merely because he said so and acted with bureaucratic officiousness have both occurred.

Negligence in theft prevention begins with negligence on the part of senior management. Indifference and apathy about controls trickle down to the lower-level supervisors. A phrase used so often in management circles as to be a ranking cliché—expect and inspect—has special relevance to the prevention and elimination of negligence in theft control.

Some negligence is related to lack of knowledge or to misunderstanding. There are firms, for example, whose managements believe that a serious liability to civil or criminal penalties necessarily exists when apprehended thieves are arrested or prosecuted but not convicted. Such beliefs lead to policies of no arrest and prosecution. In reality, the liability to losses through continued or extended thefts is far greater than that to damages. Skill and care in laying the groundwork for criminal actions will minimize any liability. It is negligent of management not to explore this question to an informed and satisfactory conclusion.

Lack of Controls

Lack of controls applies alike to physical controls, procedural controls, and management controls. Among physical controls often overlooked are those on vehicles entering and leaving premises. To illustrate: if a manufacturing plant has frequent shipments and receipts of products and materials, there is vulnerability to theft losses by leaving or placing unauthorized material in an outbound vehicle. If the only control on such movements is documentary, coupled with a perfunctory physical inspection, the theft losses will not be detected. Yet many firms have dozens of daily in and out truck movements without any assurance that material has not been loaded on or off such trucks at interior points between the entry and exit.

Lack of procedural control is typified by the release of proprietary drawings or specifications to vendors without prior screening to minimize disclosure and without attaching a legend or caution against disclosure to third persons to that which is so released. A vendor could disregard such a warning and disclose the information to another, but he would do so at his own peril. Without the warning, the sensitive data could rather easily become part of the public domain.

Often overlooked management controls with security objectives include requiring the development and publication of policy statements of admittance to premises, removal of property from premises, screening and selection of employees, salvaging or scrapping of materials, separating functions in cash flow handling, and like policies that will impose restrictions, restraints, or audit trails on items that could be stolen.

The sequence of controls development should start with the management controls to set broad guidelines. Procedural controls follow as a detailed implementation of the policies. Specific physical controls generally come as subsets or aspects of the procedures controls. In the ideal situation physical controls are the last to be selected and are dictated by the earlier management and procedure controls. In too many enterprises, the physical controls are considered first and often are not related at all to overall planning. The lack of control that results in theft losses can exist despite apparent physical countermeasures.

Target Risk Situations

Another reason for theft loss is the existence of target risks. A target risk is one that involves so much of a valuable asset as to invite attempts at theft. Some target risks cannot be avoided and require specific security countermeasures to reduce the vulnerability. Others can be prevented, and that is often the fastest, easiest, and cheapest way to avoid the loss. For example, a cash sales operation that works with a bank of $5,000 but averages daily transactions totalling only $2,000 need not expose the entire $5,000 to theft by robbery or stealthy larceny. An analysis of actual transactions (detailed even to hours of peak load) will permit less than the total bank to be withdrawn from secure storage at one time. Putting all the bank in the cash sale store would create a target risk if half or less would probably be adequate for any one day's demands.

Another target risk situation would exist if a large amount of cash or negotiables, say $100,000 or more, were stored in a single storage container. Storing at several, equally secure locations would require the thief to expend double or more the effort or risk a higher likelihood of detection to take the entire amount. Two safes, side by side in a vault, each separately locked, could reduce a sizable target risk for a small dollar investment (Figure 5 and 6).

Perhaps the single most common target risk for businesses today is the computer center or data processing headquarters. This subject was treated extensively in the preceding chapter and need not be examined again in detail. However, anything that a business can reasonably do to reduce the size of computer target risks deserves priority attention because of the usually high criticality of losses in the computer area.

Where Fraud and Theft Losses Can Occur

The short answer to the question, where will theft losses occur? is, almost anywhere that there is something that can be stolen.

The varying seriousness of theft losses will require protective measures first for target risk situations and then for the places where the likelihood of theft, even in lesser amounts, is great. The areas described in the following subsections have traditionally proved to be those of high theft loss vulnerability. The seriousness of the losses in each area will depend entirely upon the amount of assets at risk there.

The Accounting Area

Fictitious accounts payable, reused invoices, cancellation of accounts receivable, and forged or altered vouchers or expense reports are classic examples of theft techniques arising within and involving manipulation of the accounting function. Other thefts related to accounting operations can result from failure to store securely or account for forms and documents that can

Figure 5. Two-hour Fire-resistive Safe
(Courtesy of Diebold, Incorporated)

authorize the transfer of property or funds. Blank or partially completed check forms, bulk supplies of property passes, shipping authorizations, or interplant transfer forms, and other kinds of documents that can authorize the movement of property or the alteration of records regarding property require accountability and periodic verification. An appropriate form is often all that is required to process a record that can divert thousands of dollars in assets. Not only the ultimate attempt to reconcile various forms against transactions they purport to reflect but also the step-by-step control of the form itself is required.

Also related to accounting are the embezzler's practices of cash lapping and kiting. Lapping involves the transfer of funds from one cash account to another to cover unauthorized drawings or withdrawals against the latter. Kiting is the drawing of cash against questionable paper or the misappropriation of cash and the use of subsequent receipts to balance the oldest delinquent accounts. Both practices rely upon the availability of time within

Figure 6. A Tool-resistive Money Safe (Courtesy of Diebold, Incorporated)

which to make the next required maneuver. If a payment received today can be deferred on the record and the money used for personal application, then money received tomorrow, in payment of a different account, can be credited to the first account, and so on until the amount gets too big to handle or a simultaneous check of all accounts involved reveals the irregularity.

Because the embezzler generally operates from a position of trust involving not only the cash embezzled but the records concerning it, it is important to consider a number of other techniques in addition to lapping and kiting.

The embezzler can often claim a theft when none occurred, except his own, of course. He can issue checks for returned goods that were not returned and that can get lost in the accounting maze. Often the transposition of one or more digits in a manual record or the destruction of a single data card in a deck of machine records will be enough to foul up a transaction like this so badly it cannot be resolved. The embezzler can also charge items to inactive or unauthorized expense accounts to cover cash thefts. He can pad cash expenditures not supported by paid bills or vouchers. Every employee can play this game. All that's necessary is to raise the unsupported expense items a little bit. For example, a supervisor or manager is traveling on company business. He will submit paid bills or credit card receipts for his hotel, car rental, and airplane tickets. But his lunches and dinners, local telephone calls and taxis will not usually be supported by bills or receipts. If the employee is traveling one week and if he increases five breakfasts by $0.50, five lunches by $1, and five dinners by $2, he has already taken $17.50. Add one or two local phone calls and a taxi fare (when he took a bus instead) and it is fairly easy to pad (read embezzle) $20 in a week with no real probability of discovery or even suspicion. An employee who travels frequently can add to his spendable income by several thousand dollars.

The fact that paid expense reports are tax deductible by the employer and that the Internal Revenue Service demands receipts or paid bills usually only for items $25 or higher lulls many firms into acceptance of generally padded expense reporting. Aside from the actual dollar loss that this form of fraud generates, it suggests the possibility of more ambitious frauds to those who practice it successfully.

Purchasing

A major fraud practiced in purchasing is the kickback by vendors of part of the sales price to the buyer or other employee who authorized placement of the order. Kickback can take fairly subtle forms not always involving cash directly. A supplier of building materials may find himself "invited" to deliver shingling, siding, or other items to an employee's home. Theater tickets to a hit show may be received in the mail. Another popular practice is to place orders with confederates from whom invoices, but no property, will ultimately be received. Internal manipulation of receiving reports or the approval of purchase orders for payment without a review of supporting documents will provide a kitty to be divided between the fraudulent vendor and the purchasing employee confederate.

Closely allied to actual fraud in the purchasing operation is the conflict-of-interests situation in which a responsible purchasing employee can authorize or direct purchases from concerns in which he has a real (though often concealed) financial interest. There is no problem if the firm selected is competitively priced and provides quality items on schedule. But if better deals could be made, it is not likely the interested buyer will try very hard to make them.

Warehousing, Inventory, and Distribution

The opportunities in the warehousing, inventory, and distribution areas relate not only to the theft of stock but also to the use of distribution resources such as trucks, trailers, and freight cars for removal of other stolen items. Thefts on receiving and shipping docks are most common and can exist despite apparently thorough precautions. In one case, shipping dock personnel assured that each staged load was picked by one employee, transported to the loading area by another, and finally checked off by a third, all of whose counts were supposed to tally. The problem was that, in the interest of efficiency and optimizing turnaround time for common carriers, outbound loads were being checked (final-counted) before arrival of carrier vehicles, sometimes by as much as two or three hours. After the count, items were being surreptitiously added to the staged load. Since time was available and the carrier to make the pickup was known, all that was needed was collusion between the driver and an inside employee. In cases of repeat assignment of the same driver by a carrier, the problem could reach major proportions.

Many firms allow what is referred to as will-call deliveries, which are deliveries held at the delivery or shipping dock for the customer who comes in his own vehicle or sends a local transfer agent to pick up the items. Access to the shipping area by persons making will-call pickups and the use of the vehicle involved for unauthorized removal are among the most troublesome of warehouse and distribution center theft problems.

Parcel post or parcel service deliveries are another problem. In either of these forms of shipment, the employer assigns one or more employees to pick, package, address, and deposit in the delivery vehicle certain quantities of a product to be delivered through the U.S. mails or by the regional parcel service. The finished package is label-addressed according to the shipping authorization, and the amount of postage or parcel service stamps is calculated and affixed. At an appointed hour the pickup vehicle arrives and the driver verifies the total count of packages delivered to him against a manifest, assures the postage or parcel stamps are present, and leaves. In many situations there is little to prevent the assigned employee from shipping unauthorized materials to an authorized address or unauthorized materials to an unauthorized address. Cross verifications of parcels, shipping documents, postage or parcel stamps, and delivery manifests are often so delayed that losses could not be identified even though indicated.

Lack of accountability for truck and car seals is another source of theft of finished goods. The presence of a seal presumes the shipment is intact upon arrival. But seals can readily be substituted. If the supply of seals is not stored securely and if seal numbers are not accountable on each load from the moment the seal is assigned to that load until the moment it is verified upon delivery, the whole routine is illusory. Yet there are segments of the transport industry that do not maintain seal accountability as a matter of choice.

Manufacturing Operations

The most well-used technique to cover thefts or other unaccounted-for shortages from manufacturing areas is to report damaged or broken goods. When the goods are high-unit-cost items whose breakage is not easily verified, the loss can add up quickly. The TV, radio, and vacuum-tube manufacturers were regular victims of the technique. A high breakage or damage level is always an indicator of possible theft and should be audited or otherwise investigated promptly.

An allied practice involves what some persons regard not as theft at all but as adroit "oneupmanship." That is the cannibalizing or surreptitious diversion of one operation's materials or components to another's needs. In the factory it can happen as early as incoming inspection or receiving and as late as assembly operations. In the office it may involve equipment or supplies. The same lack of controls that permits this practice will, and probably does, permit theft as well.

Motor Pool or Traffic

Theft of gasoline and oil products is common. Theft of the vehicles is less common but not unknown. Unauthorized use of vehicles is widespread and may cost more than mere accelerated obsolescence if the use results from the employer's negligence or lack of control and involves personal injury or property damage.

Credit cards for fuel and services purchases are frequently used to defraud employers. The charged items are reported procured from a dealer in collusion with the employee. No sale, in fact, takes place, but a credit card charge slip is made out. The employee turns it in with his expense report or trip ticket, and the invoice is paid when received. The employee and the dealer divide the amount.

Repairs and Returned Goods

A customer finds his newly purchased appliance is defective. He notifies the vendor and is told to bring or send the item to a given address. Upon its receipt at the address, the item is booked in, verified against warranty contracts when appropriate, and evaluated as to repairability. Repairable items are repaired; the others are scrapped or disassembled for parts recovery. The decision to scrap or repair can be motivated by a theft opportunity. If good-condition products are scrapped and if the items can be removed from the returned-goods area, either by the employees assigned there or as part of a junk or scrap shipment later picked up by a regular scrap removal contractor, good items will be stolen. Several judgments on scrap decisions, surveillance of repair areas, periodic audit of scrap removal, and other procedural safeguards will reduce or prevent such losses.

There are many cases on record of the use of scrap removal resources for concealment of stolen items. In one case barrels of metal turnings and shavings were partly filled with copper casts, coils of copper wire, or other valuable products and then covered to the top with the turnings. Other variants of the same technique involve loading of higher-priced waste metal first and then covering it with lower-priced waste. Approximate corrections for weight will allow an entire truckload to be weighed and removed and later paid for by the scrap vendor on the basis of the stated weight of cheaper metal appearing on top of the barrels rather than the more expensive material buried below.

A careful analysis of the specific vulnerabilities of a given operation will reveal dozens, perhaps hundreds, of other theft opportunities. The illustrations discussed are only of common problems likely to occur in every business that does not take precautions against them. There is no substitute for the careful analysis. Even the reading of general loss prevention guides available from casualty insurance companies will not be helpful if the specific occurrence framework of the loss discussed is not identified. The checklists and guides are good sources of risk matrix design items. Evaluation of the actual risk requires the manager to study his own operation.

Prevention and Detection of Losses

First among the things that can be done to prevent and detect losses are those that should be done as a matter of general practice. They include operational as well as financial audits, development and publication of policies on the consequences of detected theft, and a separation of functions that authorize cash disbursement or materials transfer from those that review or verify the records of the transactions.

The operational audit—a review of a procedure to assure conformance with a stated instruction—requires careful development of operational procedures statements. The very earliest audit is closely akin to the vulnerability assessment. After the weaknesses are spotted and the countermeasures are established, later operational audits will aid in preventing disregard for the procedures from leading to actual losses.

A special word must be said about policies on prosecution because so many firms mistake condonation for compassion. Theft, in any amount, is a crime in every jurisdiction in the United States. It is either a misdemeanor or felony depending in part upon the amount stolen and in part upon the circumstances or mode of theft. For an employer to act as though his was the only interest at stake in the theft of his property is as wrong in terms of overall social good as for a passerby or witness to street crime to refuse to "become involved." Theft is a crime against all the people and an economic wrong against the particular victim as well. Both aspects of the theft problem must be considered.

If it is known beforehand that a firm will not prosecute a thief, the deterrence of fear of arrest and conviction is lost. Policies of nonprosecution become known even if they are not articulated in writing. Company discipline, even discharge, will not deter transient or short-term employees from theft, although it may have an effect on long-term personnel. But "no prosecution" policies are usually based on one or more of four common grounds: (1) all notoriety resulting from a prosecution will be unfavorable to the company, (2) other workers will feel the company is oppressive if it prosecutes, a situation of the "big company and the poor little worker," (3) there is a possibility of damages actions against the company for false arrest or malicious prosecution or false imprisonment, (4) dismissal from employment is enough "punishment." Each of these has a fallacy and can be eliminated by argument.

The problem of notoriety assumes no effort by the company to tell its version of criminal prosecutions. In the face of increased public concern over rising crime, accurate, prompt, and factual information releases by a company will permit the news media to publish a balanced account. If a company hedges or refuses to discuss criminal cases involving its employees, it really has only itself to blame when the press is unfriendly or publishes only information from the accused employee.

If the company believes that most workers are good citizens and are not themselves stealing, it must reason that prompt suppression of acts of theft is as much in the employees' interest as the company's. Locker, pocketbook, and desk thefts are among the most common of all industrial stealing. Employees are the victims. The company's firm stand on theft prosecution is a sign to the majority of employees that their interests are being protected too. Inclusion of this objective in the theft control policy and reiteration in the information releases will reinforce it.

The fear of damages actions is overemphasized by most companies. Although actions can be brought almost everywhere for false arrest and imprisonment or malicious prosecution, they invariably require a showing of positive malice or such negligence on the part of the company as to amount to a wanton disregard of the employee's rights. If a firm prepares its position carefully, particularly by insisting on a high level of professionalism among its protection or security employees, and if actual arrest or confinement is left to local police acting in conformance to established statute, the probability of a successful suit is remote. Certainly in cases in which the accused employee has admitted voluntarily before witnesses that he is guilty there is no concern. And even if a successful action were brought for civil damages, the company need not assume it will lose hundreds of thousands of dollars. The damages, if any, have to be proved. Reinstatement, back pay, public apology, or intervention to assist employability elsewhere if the employee wants to leave would mitigate such damages.

To argue that dismissal is enough punishment is to confound the right of the company to protect itself with the right of the state to punish crime. Dismissal is not punishment; it is prudent self-protection. A firm need not expose itself to further thefts or undertake special precautions applicable to only one employee to prevent that employee from stealing. The problem of dismissal as punishment is often further complicated by bad timing of industrial relations action. If an employee is apprehended in a theft or admits complicity and if the decision to prosecute is made but the employee is not terminated until after the trial, then failure of conviction for any of the reasons over which the company has no control will jeopardize the later company discipline.

The right to terminate a dishonest employee is not contingent upon his conviction of a crime. Even in collective bargaining situations in which discharge must be for just cause, there is no prior requirement for criminal conviction. A reasonable and convincing case (surely reasonable when the employee's own admission is on the record) is all that is required. If the company intends to prosecute as well as take internal action, the internal action should come immediately with the discovery of the offense or completion of the inquiry. The prosecution should also be initiated promptly but can proceed independently. The employee's discharge will not be admissible against him in a criminal trial (although his statement leading to discharge might be), but his acquittal in a criminal trial will surely be considered in later arbitration proceedings over his belated termination.

Some Specific Preventives

The precautions discussed to this point apply to all enterprises without distinction as to business or even type of theft vulnerability. The following suggestions are made to aid prevention of specific types of theft loss.

Cash thefts. Use lockboxes whenever possible for expected cash receipts. Supervise the opening of mail to such lockboxes

through physical and procedural controls. Insist upon accountability in cash handling: each person with unsupervised access
to cash must account for the amount before and after the access.
Reduce imprest funds, petty cash accounts, cash sales banks,
and other cash disbursement funds to the minimum consistent
with operating requirements. Store cash in suitable containers
appropriately rated for surreptitious and forced-entry prevention.

Employ intrusion and robbery alarm systems in cash handling
or storage locations. When depositing cash, use duplicate deposit
tickets with one returned directly by the bank in the mail. Separate cash receipt from cash disbursement; for example, a cashier
at a sales counter should not make refunds or adjustments. Separate cash collection from cash deposit; use different personnel
for each task.

Fraudulent manipulation of records. Countersign all checks
and drafts through an employee not involved in their preparation. Cancel paid invoices, processed receiving reports, and
closed purchase orders with a perforator or other tamper-proof
cancellation device. Use prenumbered cash and property transfer
forms and maintain accountability by blocks of numbers. Require
pen or other indelible writing on all accounting and property
records; errors can be lined and initialed, but no erasures. Separate functions of authorization, execution, recording, and review;
more than one person should be involved in every transaction
involving cash or property. Cross-train personnel and transfer
them from function to function periodically without advance
notice if possible.

Theft of property and materials. Establish property pass requirements wherever control can be maintained over entry-exit.
Institute inspection routines for inbound and outbound vehicles
and parcels. Verify all shipments against shipping documents.
Count items out of the plant or onto the vehicle by using different personnel from those who staged or picked the material.
Regard items of product or other material found in unusual
places with suspicion; such items are often cached for later removal or recovery. Register all visitors and annotate the permanent record if the visitor enters with a case or package as a
check when he departs. On a random-sample basis select out-

bound loads for tear-down examination or very close inspection. Discourage general parking of private vehicles near doors, windows, or other openings; separate parking from work areas by a fence if possible. Don't obscure warehouse or storage area windows with piled items; leave enough space between the wall and the stacks to inspect for glass breakage or removal. Distinguish but include in physical inventories and cycle counts finished goods, both in manufacturing and warehousing, work-in-process, and major components. Establish understandable break-points or cutoff dates and times for inventory counts so that numbers in various locations can be properly correlated. Attend to basic housekeeping to prevent accumulations of waste and junk that could conceal or aid removal of property.

The Undercover Investigation

Undercover investigation should always be considered among the techniques available not only for theft detection but for prevention also. It is a volatile measure and can be embarrassing, expensive, or both if not handled properly. It is generally inappropriate when flexible control over personnel is not available or when competent, professional supervision will not monitor the operation constantly.

Among the dangers of a badly handled undercover operation are possible charges of unfair labor practices by collective bargaining unit personnel, exposure and discrediting of the employer, loss of respect or affection for supervisory or management personnel in charge of the activity, and physical harm—perhaps death—for the operative or his key informants.

Among the benefits of successful undercover investigations are the exposure of concerted plans and schemes for theft activity, the identification of particular thieves and the accumulation of evidence that permits their arrest and prosecution, the discovery of procedural weaknesses that can be remedied without dramatic personnel action, and the therapeutic effect—the falling off, at least temporarily, of other thefts.

REFERENCES

1. Summary Account, American Management Association, *Crimes Against Business Project,* LEAA Grant #76-DF-99-0063

2. Task Force Report of the President's Commission on Law Enforcement and the Administration of Justice, *Crime and Its Impact—An Assesment* (Washington, D.C.: Government Printing Office, 1967), pp. 43ff.

3. *Crime Against Small Business,* a report of the Samll Business Administration, Senate Document 91-14 (Washington, D.C.: government Printing Office, 1969), P.1

4. Reported in *Industrial Security,* April 1962.

5. Note 1, supra, p. 47.

6. FBI Uniform Crime Reports show rate of non-violent larceny, *excluding fraud,* up 7% from 1975 thru 1979.

Selt Test Questions

1. What are the most common causes of fraud and theft?

2. Name five areas frequently overlooked in establishing management controls.

3. What are "will call" delivery arrangements vulnerable to theft loss?

4. What preventive techniques are useful to control losses in returned goods areas?

5. Name nine specific preventatives to reduce or eliminate cash theft.

6. What techniques will help reduce record manipulation losses?

7. What are the advantages and disadvantages of undercover investigations?

= 8 =

Guard Operations

GUARDS do not make up the whole of any security program, but most security programs include the use of guards. Guards or their more familiar predecessors, watchmen, have been the chief industrial security resource since the very widespread use of both military and civilian personnel in that function during World War II.

Dependence upon guards results in labor-intensive protection programs, which are the most expensive and are often inefficient. It should be the objective of every manager responsible for industrial security to reduce guard manpower to the lowest number of men consistent with the needs of the enterprise. Success will require the continued application of the systems techniques already mentioned to spot weaknesses and select countermeasures. Even with the most effective systems engineering, most security programs—and all large ones—will have some need for guard manpower. It is the purpose of this chapter to discuss ways to recruit and select, train, deploy, and supervise guard personnel in the most effective manner. Those considerations are particularly important as security programs become more mechanized and automated.

Although the need for manpower in absolute numbers will decrease, the requirement for greater competence among the guards who are employed will increase. Conventional manpower pools may not be appropriate for staffing guard forces of the type demanded by well-designed security systems. Most surely the training and supervisory practices will have to be modified.

Determining the Need for Guards

When are guards appropriate? The answer to that question can be stated simply this way: Guards are appropriate whenever the functions required by the security program involve the use of discriminating judgment, the application of nonprogrammatic control techniques, and the use of physical force to apprehend, direct, or restrain persons.

From that standard it is clear that guards, to be utilized efficiently, must accomplish tasks that are either too subtle or complex for machines or that involve a high degree of subjective evaluation. To illustrate: a guard is not necessary to count traffic, verify identity devices, turn lights on or off, open or close doors, or detect changes in the physical environment. Such functions are better performed by mechanical or electronic devices. However, to determine how to deal with a person who does not clearly meet the requirement for personal identification or to halt a fistfight among employees or to direct human and material resources applied to a disaster emergency, more is required than some preprogrammed machine control. By combining the machine tasks and the man tasks, a larger mix and greater number of total tasks may be accomplished. That is the objective of integrating human and machine subsystems in the protection effort.

Some of the conventional functions that are performed by guards and cannot satisfactorily be handled without them include random patrol of premises, inspection of property or material entering and leaving, issuing and retrieving permits, passes, and other documentation, reception and escort activities, fire and disaster emergency response, incident investigation, and various special assignments including driving vehicles. If those functions are present in the security program, there is an indicated need for guard manpower. The next question is, how much manpower?

Certain activities will require that the assigned guard remain in one location. Fixed-post assignments, as they are known, may require constant attendance by guard personnel; examples of such assignments are gates and entrances where pedestrian or vehicle traffic is monitored, communications centers, and control desks or consoles. It is important to minimize the number of constant-attendance fixed posts because the labor expense required to maintain such posts is the greatest of all guard expenses. For each post that requires constant manning on a 24-hour, 7-day basis, there is a minimum need for four full-time guards and part of the time of a fifth. The requirement is derived by assuming a normal workweek of 40 hours over 5 consecutive 8-hour days. Provisions for 6-day workweeks, overtime beyond 8 hours, and split weeks in which single days off are spaced among workdays will all find an application, but the basic workweek, as is typical in American industry, is 40 hours. Dividing the 168 clock hours in a week by 40 gives a quotient of 4.2. Theoretically, four full-time men and one 8-hour shift by a fifth man will provide the required guard coverage for a 24-hour post. But sickness and other emergencies cause unforeseen absences, and vacations result in planned absences. Experience recommends that each full-time guard post be manned at the level of 4.5 rather than 4.2 to provide for contingencies.

Other posts will require manpower only during certain shifts, as during the day shift at a factory gate used only during that shift. Some posts will also permit the temporary absence of the assigned guard. Typically, patrol posts are so regarded. In scheduling relief manpower, such posts need not be considered as requiring any but one man during the assigned shift, regardless of the need for meals or other absences.

Table 4 sets out four forms of representative manning charts for a guard force. By arranging the chart as shown in forms 1 and 2, the number of men required for each post during each tour is established. The next step is illustrated in form 3; it involves preliminary planning of manpower by writing in a numeral for each guard, using the standard of 40 hours, 5 consecu-

Table 4. Examples of Guard Force Planning Forms and Schedules

Form 1. Schedule of Guard Hours

Post No. and Location	Day Tour (0630–1500)		Middle Tour (1500–2330)		Night (2330–0800)	
	Reg. Hrs.	O/T Hrs.	Reg. Hrs.	O/T Hrs.	Reg. Hrs.	O/T Hrs.
1. Main Gate	8	2 (to 1700)	8	None	8	0.5 (from 2200)
2. East Gate	8	None	8	0.5 (to 2400)	Closed 2400–0630	
3. So. Gate	8	None	Patrol to 1800, then close		Closed 2400–0630	
. . .		. . .	. . .	. . .	. . .	. . .

Form 2. Post Meal Relief Schedule

Post No. and Location	Day Tour		Middle Tour		Night Tour	
	Relief	By	Relief	By	Relief	By
1. Main G.	0930–1000	Patrol A	1800–1830	Patrol A	0230–0300	Self
2. East G.	1015–1045	Patrol A	1845–1915	Patrol A	Closed	–
3. So. G.	1100–1130	Sgt.	1915–1945	Sgt.	Closed	–
4. Patrol A	1045–1115	Self	1915–1945	Self	0300–0330	Self
. . .			. . .		. . .	

Form 3. Personnel Assignment by Post, Tour, and Day

Post	Day (0630–1500)							Middle (1500–2230)							Night (2330–0800)						
	S	M	T	W	T	F	S	S	M	T	W	T	F	S	S	M	T	W	T	F	S
1.	–	7	1	1	1	1	2	–	8	8	8	8	8	–	–	–	–	–	–	–	–
2.	2	2	2	2	3	3	3	9	9	9	9	9	11	11	15	15	15	15	15	16	16
3.	–	4	3	3	4	4	–	–	*10	10	10	10	10	–	–	–	–	–	–	–	–
4.	5	5	5	5	5	6	4	11	11	11	12	12	12	12	16	16	16	17	17	17	17
5.	4	–	–	–	7	–	7	12	*10	10	10	10	10	14	17	18	18	18	18	18	19
6.	–	6	6	6	6	7	–	–	13	13	13	13	13	–	–	–	–	–	–	–	–
7.	7	–	–	–	–	–	1	–	–	–	–	–	–	–	19	19	19	19	20	20	20
			. . .							. . .							. . .				

* Posts 3 and 5 are split on middle tour, half on one and half on the other.

Form 4. Days off Schedule

Guard	Days off	Guard	Days off	Guard	Days off
Jones	Mon, Tue	Black	Sat, Sun	Name	Wed, Thu
Smith	Tue, Wed	Green	Wed, Thu	Name	Sun, Mon
*White	*Thu, Sat	Brown	Mon, Tue	. . .	. . .

* Problem schedule: days off not consecutive.

tive days. That avoids the frequent confusion and wasted hours caused by using actual names in a schedule and constantly backtracking to be sure the schedule meets the standard. When the manning has been carried through by using the process, there will usually remain shifts and posts for which either no man is available or hours in scheduled workweeks of specific guards are unapplied. By revising the schedule to fill the requirements without violating the standard, the irreducible number of broken or incomplete workweeks will be determined. Form 4, made up after form 3 is complete, is a double check on consecutive days off. Depending upon the total number of hours involved, the requirements can be met by scheduled overtime or by adding men who will be supernumeraries on one or more days. The final decision will be made on a combination of cost and overall protection needs. When no task really useful to the protection mission would be accomplished by extra manpower, overtime will be used. As a rule, the planning should assume overtime as the solution.

Peak Demands on Manpower

It is usual to find guard force schedules involving consecutive 8-hour shifts. A typical day would have three shifts: 8:00 A.M. to 4:00 P.M., 4:00 P.M. to midnight, and midnight to the following 8:00 A.M. Equally often the shift will not contain a scheduled meal period. Each man will eat on post or when time is available. Theoretically, each man remains on duty through such meals. Justification for the practice is that the least expensive schedule is developed without a need for relief manpower to accommodate formal meal periods. In some situations the approach is effective. However, there are many cases in which overtime is otherwise required at critical times of the day because of peak movements or activities in the plant, factory, or office, and the costs of such overtime can equal or exceed the presumed savings derived from nonscheduled meals.

A more flexible approach involves preplanning the day to determine when the peaks occur and then establishing regular

guard shifts to overlap such peaks by adding to them the scheduled meal period. A 30- or 60-minute meal period will result
in 8.5 or 9 working hours and will permit the use of those extra
hours for shift overlap at no premium in overtime. Figure 7 shows
a planning graph for the application of the technique. The overlap approach does not require that time be treated in equal segments. A given application might suggest that during two
1-hour shift-break periods, there be double the normal manpower
available and normal manning at other times. By assigning one
shift from midnight until 9:00 A.M. with a 1-hour meal, a second
shift from 8:00 A.M. until 5:00 P.M., also with a 1-hour meal,
and the third shift from 3:00 P.M. until midnight, with a 1-hour
meal, double manpower will be available at no cost increase
during the periods 8:00 A.M. to 9:00 A.M. and 3:00 P.M. to 5:00
P.M. The scheduling can be arranged in many ways to obtain
full or fractional hour overlapping.

One obvious requirement is that provision be made for the
relieved meal. It can be made by assigning meal periods during
the lightest part of the guard shift involved and by providing
for sufficient manpower on a mobile or nonfixed post basis to
accommodate required fixed-post relief. Often, temporary
changes in conditions at a post will permit the adjustment. With
appropriate notification and planning, a gate normally open and
manned for employee ingress and egress can either be handled
on a remote basis through the kinds of controls discussed at
length in the later chapter on systems or be closed for an otherwise light hour or part of an hour and thereby free the assigned
guard to relieve another post. Meal and other necessary relief
provisions may also be made through the temporary use of non-
guard personnel, as at reception desks.

Sources of Guard Manpower

Two ways to supply needed guard manpower are possible.
One requires that an employed guard force be developed and
administered. The other permits the use of contract or guard
agency personnel. Some situations can best be handled by a com-

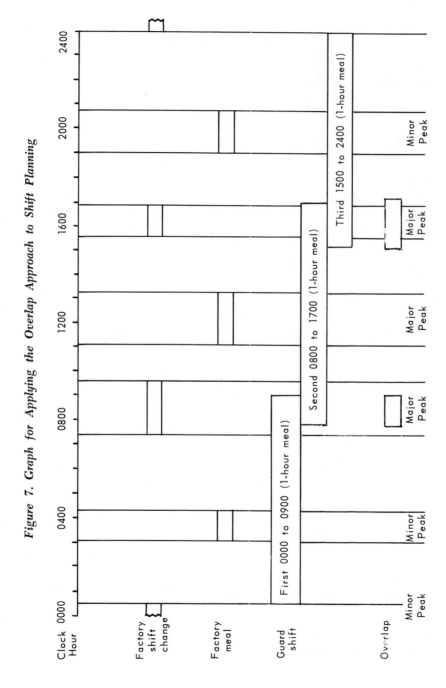

Figure 7. Graph for Applying the Overlap Approach to Shift Planning

bination of both techniques. Each technique has distinctive advantages and disadvantages, and the decision to use one or the other can be made properly only if the characteristics of each approach are fully understood and evaluated.

The Employed Guard Force

The first important characteristic about an employed guard force is that the continuing requirement for maintaining regular and relief manpower remains with the employing organization. All the personnel will be on the enterprise payroll. Although some different treatment among various classes of employees, including guards, will be possible, basic considerations such as vacation, pension, sick absence allowance, and other employee benefits will probably be extended to guards. The cost of employed guards, then, will involve all the fringe expenses in addition to the direct labor expense. If guard forces are large, that can be a major item of expense. For example, if the paid hourly guard rate in company X is $3, the average cost of benefits is 20 percent of the base rate, and the applied overhead is $1 per productive hour, then an 8-hour day will cost $36.80, not $24.00. Some tangible benefits should accrue to justify expenditure of the additional $12.80, because the labor could be obtained at the basic rate through the use of an agency, at typically lower cost.

The benefits alleged for employed guard forces are usually these: loyalty to the enterprise through identification with it, continuity of service, greater latitude in personnel selection, and better opportunity for cross-training. Those benefits are generally easier to achieve with an employed force rather than an agency force, but it does not follow as a matter of course that they will accrue merely because the guards are employees. Appropriate use of management controls with respect to agency personnel can produce good results too. The kinds of controls needed are discussed further on in the chapter.

The disadvantages of employed guards, in addition to the cost factor, include superannuation, particularly when no compulsory retirement policy exists, and the requirement to provide for vacation and illness by adding personnel beyond the actual manning needs. There may also be collective bargaining contract obligations when members of the guard force are part of a bargaining unit. Bargaining contracts usually include provisions relative to just-cause discharges, prior benefits clauses, and arbitration provisions. Although they are unquestionably of value both to the worker and the employer in an overall sense to prevent arbitrary conduct that deprives a worker of job or income, such contract commitments can be a special problem in security. For example, the most fundamental concept in bargaining is seniority. However, application of seniority provisions to guard operations has produced very bad results by freezing personnel to particular posts. If every transfer of a guard must be based either on seniority or on some provable disciplinary requirement subject to the grievance and arbitration process, guard personnel will have a more successful hold on specific post assignments than is good for sound protection. Rotation of assignments is important to security work because it limits the predictability of future duties, which in turn reduces the likelihood that improper arrangements will be made or continued among guards and other employees. In addition, the value of a guard force as an emergency response resource demands that each member be as familiar with the functions of other members as possible. The only practical way to assure that is by rotating assignments.

The factors that will help determine whether to use an employed guard force rather than an agency are these: (1) the availability of qualified manpower both to the enterprise and to the agencies, (2) the size of the guard force (the larger the force the more efficiently employed manpower can be used), (3) the amount and quality of supervision to be provided for the guard force, and (4) the nature of the guard force duties.

When the number of guards is significant (20 or more), the duties are broadly varied or complex, the area labor supply is limited either by a lack of manpower in general or the absorption of available manpower into other jobs, and flexibility and freedom of movement is possible in administering an employed guard force, serious consideration should be given to using that approach. Conversely, when the number of men is small, the duties are fairly easy to learn and perform, the manpower pool is large, or the flexible management of an employed force is seriously restricted by policy or bargaining considerations, planned use of an agency is indicated. Whether the guards are employed by the enterprise or by a guard agency, the training and supervisory techniques explained later must be properly used.

The Contract Guard Force

Contract guards are provided by two types of company: the national organization operating through local branch or district offices and the purely local company that may serve only the particular community. There is little to suggest that the national company is a better choice merely because of greater familiarity with it through advertising, publicity, and contract assignments in many locations. The important standard for determining which agency to use is local reputation and record. Except for the emergency reassignment of personnel from other locations under extraordinary circumstances and for meeting the requirements of large clients, a national organization will not contribute much to the success an enterprise has with the locally hired and assigned guards. Local organizations frequently have advantages because they were founded by local citizens whose ability to relate to local government agencies and other community groups is greater than that of out-of-state or national companies.

In any event, the guards assigned will surely be local people, the rates paid will have to compete in the local market, and the operational requirements will be more affected by local laws and regulations than any other. When an agency is selected, agencies of both types should be invited to respond to specifications. The care with which the specifications are drawn and the willingness of the agency to bid to them will be a more reliable indicator of the agency's potential value than any other.

The specifications should include those relative to the individual guard and his qualifications, those relative to the task or guard assignment and the requirements for its satisfactory execution, and those relative to the agency as a source of trained manpower and supervisory resource.

Specifications for Contract Agencies

The first consideration is the man or guard specification. In this regard three items are of paramount importance: the guard's physical condition and general physical ability, his mental or intellectual ability, and his character or ability to meet criteria of personal habits, moral values, and rules of conduct. Table 5 shows a list of suggested guard specifications that reflect minimum standards in this regard. It is important to secure a response from a guard agency that it has met or will meet the specific criteria rather than a general statement that agency personnel are investigated by the agency or that they meet the licensure requirements of the state. Frequently licensure requirements demand only that a guard agency obtain a completed history state-

Table 5. Recruitment Standards for Guard Force Personnel

Physical Condition
Age between 21 and 50 years
Weight proportionate to height
No impairments of hands or feet
Vision correctable with glasses to read all normal correspondence and materials without further magnification
No communicable disease
Prehire medical exam with results satisfactory under existing client standards

Literacy and Education
Functional literacy or better
Primary education completed
Secondary education completed highly desirable
Ability to pass screening battery

Character and Reputation
No criminal record or history of bad character
Able to procure a weapons permit
No history of financial irresponsibility or unfavorable litigation record
Background investigation completed prior to employment and results available to client

Special Skills
State driver's license
Bondable under fidelity coverage

---

ment and fingerprints from guard job applicants. There is usually no requirement that the history statement be investigated. If a file check with the state's criminal records agency reveals no criminal record for the applicant whose fingerprints were submitted, employment by the agency will be lawful. In such cases the agency may know nothing and never learn until too late of noncriminal problems such as alcoholism, financial instability, mental illness, or unsatisfactory work record with prior employers.

Note that the specifications suggest an age range. That is to assure sufficient maturity at the younger end of the scale and to avoid the onset of physical debility at the older end. Federal and some state laws prohibit discrimination by an employer on the grounds of age alone. Some very specific reasons must be alleged to deny consideration for employment merely because of the applicant's age. The restraints would also operate upon the enterprise if it should recruit and employ its own guards. However, to suggest a recruitment objective to a contract agency to be met by the agency through selective assignment rather than discriminatory hiring will relieve the enterprise of direct compliance responsibility and still permit some discretion in personnel selection.

As a further assurance that the personnel assigned by the agency meet the specifications, the agency can be required to submit evidence of the background investigation it conducted, and final approval of the assignment can be contingent on the results' being within the limits of the specification. That is a practical way to assure agency compliance with the man specification.

The next specification is that of the job. Because post duties will change over time and assigned personnel should be capable of performing all the duties of all the posts, it is better to draw a general specification including the range of tasks that may be represented in all the posts. In addition, the training program will be responsive to specific post requirements; thus it will cover both the original selection of manpower capable of doing the job and the preparation of each assigned man for the job at hand. Table 6 suggests the common tasks that should be included in a job specification for a typical industrial guard. Tasks that are peculiar to the enterprise can be added as needed.

The last set of specifications in Table 6 refers to the guard agency itself. Some of the suggestions made here are new in the field of contract guard service, and some contract agencies

Table 6. Guard Job Specifications

Access Control

Gatehouse activity
Permit issuance
Receptionists and escorts
Employee and visitor badging
Key and lock control

Patrol

Watch clock rounds
Buildings and ground patrols
Fire housekeeping inspections
Parking and traffic controls
Lost and found

Fire Safety

Fire systems and equipment inspections
Fire fighting
Volunteer squad training
Emergency and disaster activities

Plant Protection

Internal investigations
Property controls
Enforcement of laws and regulations

Miscellaneous

Classified defense contract security
Operation and maintenance of automotive and specialized
 security equipment
Communication coordination
Insurance carrier liaison
Police liaison
Ambulance service

Agency Requirements

Turnover control (premiums rates; incentive pricing
Adequate supervisory practices)

---

will be unwilling or unable to meet the demands. To the extent
necessary, some compromise solution may be achieved by modi-
fying the specifications. The larger the account, measured in
billable guard hours, the greater the incentive to the agency.

To achieve the greatest advantage, it may be in the interests
of larger companies that have multiple plants or sites operating
on a purely decentralized basis, each of which works out its
own guard agency arrangement, to combine the requirements
within practical area boundaries. For example, a firm with two
or more sites within a single county or contiguous counties of
a single state may find it easier to negotiate a satisfactory agency
contract if all requirements are combined. Local autonomy of
plant management need not be sacrificed, because local job specifications can remain distinctive.

The first suggested agency specification deals with control
of turnover, which is the single greatest problem with contract
guard agencies. In some situations it reaches several hundred
percent on an annual basis. It may be due to the difficulty of
retaining manpower, the client's lack of objection, or both. Turnover militates against efficiency because it wipes out training
efforts. Whatever the agency or the client has done to prepare
a particular man for an assignment, loss of the man cancels out
all benefits. The difficulty of retaining manpower is the most
common reason for turnover, but there are available solutions.

Typically, a guard agency assigns 70 cents of each dollar to
direct labor, 20 to 25 cents to administrative and general expense,
and the remaining 5 or 10 cents to profit. Characteristically, the
low profit margin demands assured minimum manpower scheduling to attract any guard agency to the business. A single client
with large enough needs, however, can provide the incentive.
The objective of the agency, of course, is to hold or increase
the profit margin. When recruitment problems loom, agencies
often suggest to clients that increased rates will attract better
men and hence improve stability of the force. For that to be
true, however, the increase in direct labor must be enough to
enlarge the labor market from which the agency draws. An increase of 5 cents an hour or $2 a week is not likely to attract
new candidates into the labor pool, but 20 cents an hour or $8

a week might, and 50 cents an hour or $20 a week surely would. But the formula stated earlier says that for every 70 cents in direct labor, there will be 30 cents in agency costs and profit. To pay an additional 50 cents to guards, the agency will want to bill the client 71 cents. The upward escalation makes the cost savings originally anticipated unavailable and eliminates or lessens a primary reason for use of the agency.

If it is agreed that additional money will be allocated only to the direct labor and related payroll tax expense, the total cost to the client can be held down, the agency profit margin can be maintained, and the better manpower can be attracted. If agency manpower is not satisfactory to begin with, there is no reason to increase agency profit merely because acceptable manpower is later provided. An arrangement limiting premiums to direct labor and taxes will provide a basis upon which the agency can actually improve performance with the better manpower. At suitable intervals the profit can be renegotiated as well.

A second approach to turnover control is target pricing of the contract. That involves agreement between the agency and the enterprise on what would constitute an acceptable turnover level. It can be derived in part from the Bureau of Labor Statistics data for like jobs in the area, in part from the enterprise's experience, and in part from the agency's experience. The important thing is that the turnover level agreed to represents an acceptable rate in terms of guard force efficiency and an achievable target.

Once agreed to, incentive pricing can be attached to the rate. For some stated period, say one month or one quarter, the actual turnover can be measured. To the extent it was greater than the target, the billings for that period or the preceding period could be reduced by that percent. To the extent that it was less than the target, the billings could be increased. If the rate is realistic, performance either over or under target should balance out. Serious effort by the agency to control turnover would produce a net increase in profit, the more valuable because not diluted by direct labor or administrative factors.

Disregard of the problem by the agency would threaten not only continuance of the account but loss of booked income as well. Most agencies operate on a 30-day termination provision, and termination of the account has been the only negative inducement in the past. The agencies are enabled to remain in business by account swapping brought about by similar cancellations or terminations affecting competing agencies, but the net benefit to the client is not enhanced. If guard agencies serve a valuable purpose, then the value should be optimized for both the client and the agency. Target or incentive pricing will help.

Supervising the Guard Force

Effective supervision of guards is the single most reliable control on guard performance. Supervision includes observation and inspection of guard performance, testing of guard capability through determination of training material assimilated, counseling on and correction of unacceptable or inefficient work practices, and continuing appraisal of individual manpower effectiveness for purposes of promotional development or elimination.

It is indispensable to achievement of effective supervision that the supervisor see and talk to the guards during periods of duty assignment. Such an obvious proposition should not require further comment, but it is a fact that supervisors are often either themselves assigned to fixed posts that obviously prevent their free circulation or assigned on a visiting basis only and are unavailable when work situations that require corrective attention arise. The first practice is not unusual among companies that employ their own guard personnel; the second is a fault of agencies.

Although a guard is frequently called upon to perform tasks and make decisions that require greater discretion than the lowest-level factory hand must exercise, the guard is still a first-level employee. If his decisions are more complex than those of first-level factory hands, it is because the job of an industrial guard is characterized by more variables. In short, it requires a higher level of entry level skill. But it also requires supervision. The mistake often made regarding guards is that because they can perform at a higher discretionary level, they are capable of performing without supervision. That has frequently led to poor discretionary response by guards, development of nonstandard routines, and failure to enforce company policy uniformly.

Supervision must be distinguished from management control. A line or staff manager may be responsible for the overall performance of the guards without having the time or specialized skill to provide the hour-to-hour supervision the guards require. The usual approach, particularly with larger guard forces, is to provide shift supervisors who, in turn, report to overall guard force supervisors. Agency guards have either an assigned supervisor or a visiting supervisor. The visiting supervisor is not a reliable technique, particularly if the identity of the visiting supervisor changes.

Whether the supervisor is agency- or company-employed, the man designated for that function must meet the following tests if he is to be effective: (1) He must be of superior skill and greater experience than the average guard. (2) He must have the authority to take immediate personnel action in respect to the guard force. (3) He must visit each guard and guard post on a number of occasions during each shift. (4) He must be equipped with some objective techniques for measuring and evaluating guard performance. (5) His status and authority must be reasonably supported by facility management.

Promotion of superior personnel within the guard force in companies having their own forces will provide the skill and experience levels required. Insistence when contracting with the agency for like differentials and requiring the agency to document them will provide for supervisory skill from that source. A frequent practice of guard agencies is to accord senior rank such as sergeant or even lieutenant and captain to men of more senior tenure as a way to increase their compensation. However, the mere presence of senior rank does not imply supervisory skill. To avoid any problem of the extent of authority, true supervisors should meet the standard definitions of the fair labor standards and federal wage and hour laws; that is, they should have and exercise the discretionary authority required. Lead men can supplement supervisors, and often a shift lead man will be designated by some intermediate rank such as corporal or sergeant. In small forces in which fairly stable manpower is available and satisfactory performance history has been achieved, lead men can be sufficient on some shifts. Large or complex or highly variable shift situations, however, require supervisory oversight to optimize performance.

Visitation of Guard Posts

Actual visitation of guard posts is the key element in the whole supervisory function. Practices that grow up among guards will remain unknown to management unless there are supervisory visits. For example, a guard will often substitute personal recognition of employees, particularly managerial employees, for the requirement that prescribed identification be displayed. The longer the practice persists, the more difficult it will be to dispel the unfavorable effects. Another example is a cursory check of outbound materials and property instead of careful inspection and documentation. The most typical problem of unsupervised guards is general inattention to duty that sometimes leads to abandonment of the post or sleeping. Unscheduled supervisory visits will reduce these problems.

But much more can be accomplished by supervisory visits than mere elimination of bad practice. Positive good practice can be reinforced. All guard assignments require some instructions. The better instructions are written, organized in some logical fashion by topic, amply illustrated by actual items or photographic reproductions of items mentioned, such as identification devices, passes, and permits, and regularly updated. The supervisor will enhance an already good system of guard force instructions by checking the presence and condition of the instructional material and by engaging in short dialogues to test or review the guard's knowledge of them.

Training capsules can be developed to be presented daily and reinforced through one- or two-minute dialogues between the supervisor and the post guard.

In cases utilizing the formation or general assembly of guards preparatory to reporting to posts, the training capsule can be presented by the shift supervisor in charge of the formation.

Thereafter, the supervisor visiting the posts can elaborate it, differently, with each guard. A supervisor's report at the end of the shift would establish a record of the training accomplished. That technique alone, in the course of a year, can permit substantial review on a face-to-face basis of the entire body of guard force instructions.

Preparation of the basic guard procedures and also of the periodic or capsule training materials by a responsible manager for implementation by the guard supervisor will prevent one other common problem from getting out of hand. That problem is the growth of wrong policy or procedure, and it arises when the only training is on-the-job training by the supervisor. Supervisors can and do make mistakes, both of interpretation and application. If they are the only source of training for the guards, the mistakes will be perpetuated. No matter how small the guard force, a responsible enterprise manager should play a substantive role in the development and dissemination of policy and training materials.

Guard Administration and Reporting

Guards make more observations of conditions and circumstances that affect enterprise security than any other class of employee. Not every observation is reported, and many that ought to be reported remain known only to the guard concerned because the system for collecting such data is defective or difficult for the guard to use. Typically, narrative reports either of incidents or of whole shifts are required of guards. Some narrative reports, as for example the entries made in a continuing log, are essential and may be the primary records source for certain kinds of information. Other data, however, such as defective lights, fences in disrepair, machinery left running, doors not closing properly, fire-hazardous materials, and the dozens of other items a typical guard will observe in the course of a shift, need not be reported at length. They need only to be identified and the information passed along to some person responsible for taking corrective action.

Some interesting developments have occurred in the area of guard force reporting to improve the likelihood that the bits of vital information will be passed on for action. The techniques of machine records processing have been applied and the data have not only been collected but collected in a format that permits various statistical management reports not economically possible in manual reporting systems. The approach is simple and could be tried by many organizations that are still relying on manual records.

The first step is to compile a frequency chart of the kinds of events that, historically, have occurred with some regularity at the location. The source of the information will be past reports of the security department or guard force, reports of the maintenance and building services organization, fire insurance inspection and loss reports, and other available cumulations of information regarding security-type incidents. The purpose of the step is to identify the kinds of security-oriented information that have a frequency rate. Isolated incidents will also be identified, but they may not be included in the machine-processing phase of basic record gathering.

When the incidents have been identified, a comprehensive plant sectoring or area identification system will be required. That may already exist through column and row markings or similar grid techniques for localizing interior areas. The objective is a system that will permit fairly specific identification of plant locations with a minimum of verbal or symbolic data. A coordinate such as F-9, for example, might indicate the plant area where row F (vertical line) meets column 9 (horizontal line).

Next, the automatic data processing systems organization can help by designing a data card with adequate fields to report the following information: (1) incident, by code from the frequency list, (2) location, from the locator system, (3) date and time of occurrence, (4) department or function involved, from the chart of accounts or other internal departmental numbering scheme, and (5) identity of the reporting guard. Supplies of such machine record cards can be designed and prepared by utilizing the mark-sensing capability of existing automatic data processing equipment.

Many medium-size and large organizations are already using mark sensing and have the necessary readers. Basically, the readers are optical sensing devices that detect markings by special pencil in certain data fields on a standard data card, translate the presence of such markings to another data storage medium such as punched cards, paper tape, or magnetic tape according to programmed logic, and thus permit the retrieval of the reported data in a variety of formats.

Cards that already bear the identification of individual guards can be prepared. Supplies of such cards are then used by the guard in the course of his tour of duty to report individual incidents by the proper markings in the appropriate fields. Incidents for which no machine processing has been planned (the isolated or extraordinary incident) can be accommodated by a section for narrative comment. As each incident is reported, the guard merely moves the data card into a pouch or envelope until the end of the shift or some earlier time when the shift supervisor can collect the cards. Of course, problems requiring immediate action and response are not treated in that way. The reporting is not a substitute for guard action. It is a convenient way to record observations and actions.

At the end of the shift the cards supplied by all the guards will be examined by either the shift supervisor or the succeeding shift supervisor. He will extract data for immediate action, assure that all cards are complete, initial or sign them, and send them on to the processing station.

At the processing station the information will be converted by the sensing readers and made available for management reporting. Thereafter, at stated intervals, the information can be retrieved for review and action. Some of the most useful applications of these data are reports by type of incident, location and type of incident, department and type of incident, frequency of incident, and identity of reporting guard and numbers of reported incidents. Such reports can provide a line manager, for example, with a basis for taking supervisory and management action on events not noticed before as being problems. Doors unlocked, windows open, machines running, and areas not cleaned can be dealt with in the light of specific information.

That will contribute not only to a more secure plant or office
but to a generally more efficient one. Figures 8 and 9 illustrate
a typical mark-sense card and report page.

As such an approach develops, traditional manual reporting
may be discontinued or abbreviated in other areas as well. It
is a commonplace in security operations that one of the most
difficult tasks is to assure timely reporting by patrol- or guard-
level personnel. Possibly because of temperament, education, language skills, or self-image, guards are reluctant to reduce observations to written reports. When they do, the reports often con-
sume more time in their preparation than the contents justify,
and they almost always require more time to read and dispose
of than is necessary. The basic information is most important,
but it is missed or buried in a sea of words under conventional
reporting systems. For organizations with even first-generation
ADP equipment, the change is possible. For use with more advanced data management facilities or in companies with large
and complex security systems requirements, more fully automated techniques for data gathering and reporting are outlined
in Chapter 10, "The Systems Approach to Security."

The most fruitful application for the technique would be with
large guard agencies whose broad exposure would permit statistical profile development of great predictive value.

The Importance of Patrol

It was pointed out earlier that the most expensive and least
effective guard utilization is on fixed-post assignment. Patrol, the
opposite application, is the most effective employment if the
patrol is carefully designed.

Until fairly recently and even now in some places, the main
purpose of a guard patrol has been to prevent or extinguish fires.
The development of guard clocks and other rounds supervision
devices was motivated principally by fire insurance carriers who
insisted on some physical evidence that a guard was making
internal fire-prevention patrols. That function is still important,
although it is much less so with the development of highly reliable fire-sensing and detection systems. However, the routine

Figure 8. Mark-Sense Data Card

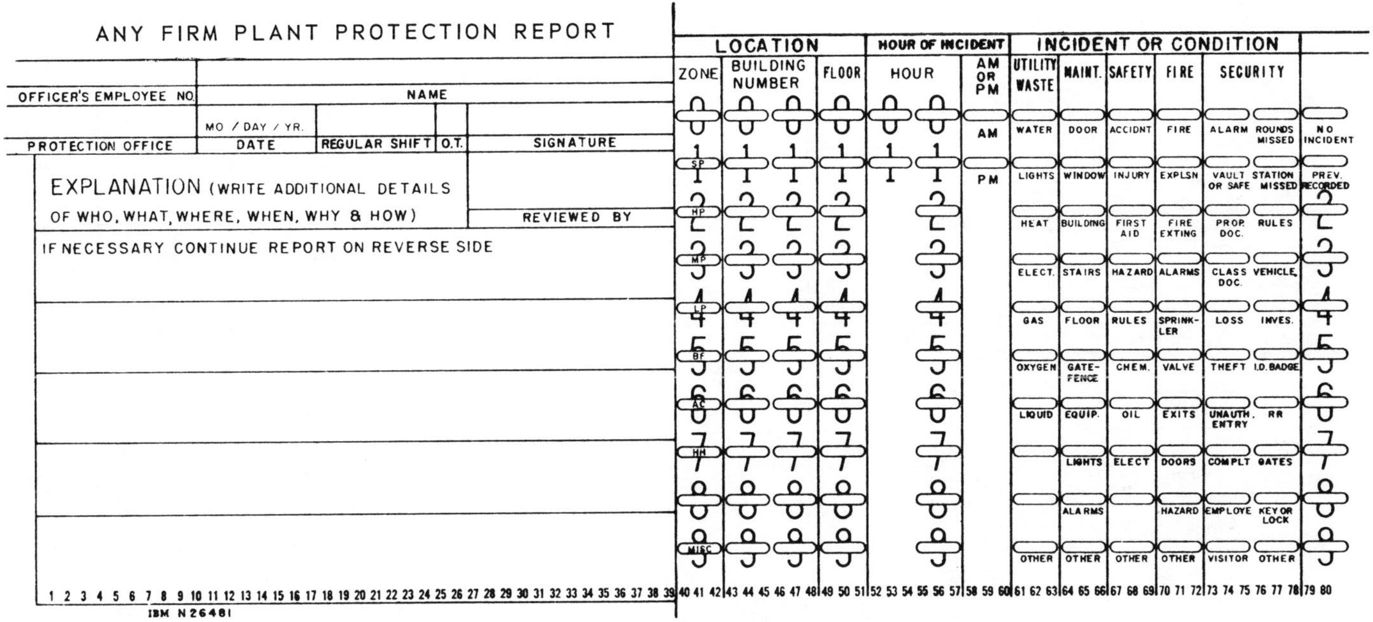

The card is prepunched with the name or shield number of the guard who will use it and is issued to that guard in packets that may be carried in a plastic carrying case similar to that used for checkbooks. With a special pencil that is provided with the cards, the guard marks a card for each incident he is reporting. After analysis by the supervisor and extraction of data for immediate action, the card is sent to the data unit, where it is processed through appropriate unit records equipment.

Figure 9. Mark-Sense Status Report

TO _____
FROM _____ SECURITY DEPARTMENT

ANY FIRM

PLANT PROTECTION SUMMARY REPORT

*PREVIOUSLY REPORTED

ZONE	BLDG. NO.	FLOOR	DATE	HOUR	UT. WASTE	MAINT.	SAFETY	FIRE	SECURITY A	SECURITY B	*	INCIDENT NUMBER	REMARKS
9	1	1	1/21/81	4PM			2					33	R.P. NORTHRUP / F CLARK FIRST AID
9	1	1	1/21/81	5PM			2					34	R.P. NORTHRUP / J SAAR FIRST AID
9	1	1	1/25/81	5PM					0			35	R.P. NORTHRUP / LATCH BROKEN DRS BETW BLDGS
9	1	1	1/22/81	6PM					0			36	R.P. NORTHRUP / LOAD DOCK NOT ALARM
9	1	1	1/23/81	7PM		9						37	R.H. RICE / SWITCH OFF PRESS-CHAM TEST RM
9	1	1	1/28/81	7PM					2			38	R.P. NORTHRUP / W B MORAN PROP
9	1	1	1/25/81	8PM				9				39	R.P. NORTHRUP / CHECKED PUMP HOUSE
9	1	1	1/26/81	8PM		6						40	R.P. NORTHRUP / CHECKED PUMP HOUSE
9	1	1	1/27/81	8PM			2					41	R.P. NORTHRUP / R ROBINSON FIRST AID
9	1	1	1/23/81	9PM					0			42	R.H. RICE / AMCO SERVICE MAN SET OFF ALAR
9	1	1	1/27/81	9PM		6						43	R.P. NORTHRUP / CHECKED PUMP HOUSE
9	1	1	1/21/81	10PM			2					44	R.P. NORTHRUP / S GARRSI FIRST AID
9	1	1	1/28/81	10PM		6						45	R.P. NORTHRUP / CHECKED PUMP HOUSE
9	1	1	1/21/81	11PM		9						46	R.P. NORTHRUP / CHECKED PUMP HOUSE
				TOTAL	0001	0012	0020	0001	0004	0008			

CODES

UTILITY WASTE
0. WATER
1. HEAT
2. ELECT.
3. GAS
5. OXYGEN
7. LIQUID
9. OTHER

MAINTENANCE
0. DOOR
1. BUILDING
2. STAIRS
3. FLOOR
5. GATE-FENCE
6. EQUIPMENT
7. LIGHTS
8. ALARMS
9. OTHER

SAFETY
0. ACCIDENT
1. INJURY
2. FIRST AID
3. HAZARD
4. RULES
5. CHEMICALS
6. OIL
7. ELECT.
9. OTHER

FIRE
0. FIRE
1. EXPLOSION
2. FIRE EXTING.
3. ALARMS
4. SPRINKLER
5. VALVE
6. EXITS
7. DOORS
9. OTHER

SECURITY A
0. ALARM
1. VAULT OR SAFE
2. PROP. DOCUMENT.
3. CLASS DOCUMNT.
4. LOSS
5. THEFT
6. UNAUTH. ENTRY
7. COMPLAINT
8. EMPLOYEE
9. VISITOR

SECURITY B
0. ROUNDS MISSED
1. STATION MISSED
2. RULES
3. VEHICLE
4. INVESTIGATION
5. RD. BADGE
6. RD.
7. GATES
8. KEY OR LOCK
9. OTHER

deployment of a guard on recurring, predictable patrol rounds makes guard usefulness minimal for crime prevention and assets protection tasks other than from fire loss.

If maximum usefulness is to be derived from guard patrol, three principles should be observed. First, the patrol route should be selected on a random basis whenever possible. Second, the patrol frequency should be directly proportional to the vulnerabilities in the area. Third, there should be constant communication between the patrol guard and his headquarters or supervisor and a capability to redirect or relocate the guard immediately.

Random patrol can be achieved by designating areas and frequencies in shift assignment instructions without prescribing the fixed route. Even if traditional clock stations are used, the sequence in which they are visited need not be fixed. What is important is frequency. Under advanced systems control utilizing digital computers, it is possible for the computer to predict those clock, key, or other reporting stations from which a guard could next report and choose any one of the available routes open after the last report. Failure to report or a report from a station not logically within the patrol route would produce an exception incident or require some explanation and would create a record for subsequent review by management.

The communications capability can be provided by radio transceivers. Reliance upon plant telephones along the patrol route is unsatisfactory because the patrolling guard is unavailable while he is between telephones. Also, if a guard should be injured or become ill while he is alone on patrol, he may be unable to reach a telephone. Radio transceivers can keep the guard in constant touch with headquarters. Every incident handled or response made by a patrol guard should be preceded by a radio message indicating the location and the problem. Failure to reestablish communication after a reasonable interval would require dispatch of additional manpower to the location last reported.

Even in situations in which only one or two men are working, the radio report is effective. When two or more are assigned to the same facility, they can report to each other. When only one man is assigned to a facility, an arrangement can be made with another facility or, if the guard is an agency guard, with the agency office for report coverage. Frequently a patrol guard will observe a crime in commission or an incipient fire or other serious emergency. Possession on his person of a radio transceiver will enable him to summon appropriate aid without loss of time and without excessive risk of injury to himself or loss of his life.

Firearms and Deadly Force

A frequent question asked about industrial guard operations is whether the guards should be armed. The best answer seems to be *no* unless certain specific criteria exist: (1) The guard will be exposed to probable attack or threat against him by persons offering deadly force. (2) Persons whom the guard has a duty to protect will probably be exposed to attack or threat against them by persons offering deadly force. (3) The guard, for good reasons, must discharge the duties of a police or peace officer as well as industrial guard.

To deal with the first criterion, it should be noted that the typical industrial or commercial operation does not generally offer the threat of deadly force to a guard or other employee. Although incidents involving deadly force will occur from time to time, particularly if target risks exist, industrial complexes or commercial sites are not the customary places for violent crime. If the guard is not a police or peace officer and if the threat of deadly force is absent, there are no compelling reasons for possessing weapons. It is sometimes argued that the possession of a deadly weapon by a guard creates a psychological advantage. That, of course, will be true only if the people in respect to whom the presumed advantage exists believe that the guard will or may use the deadly weapon. If such persons are fellow employees, business visitors, and others having permissible or even necessary relations with the enterprise, the alleged advantage is in reality a disadvantage because it encourages resentment against the guard.

The guard is not a police officer or a soldier, and his mission is not to take or, generally, even defend life. He is an extension of the enterprise management whose function is to protect and conserve the enterprise assets. Under usual circumstances when situations that involve threats to life or otherwise require response with fatal force arise, the municipal police can and should be summoned. Unless the circumstances fall within certain exception situations, deadly weapons are not appropriate for the industrial guard force.

The exceptions to the rule are not many. First, if there are target risks that will attract or invite criminal attempts, then failure to arm guards while such risks are present may place the guards and others in greater danger than use of the weapons. An example will clarify this exception. Some industrial firms pay all or some employees in cash. There may be legal or policy reasons why such a practice cannot be changed. If the amount of cash is large, the presence of the payroll will be an inducement to robbery. The presence of unarmed guards will not deter armed robbers. The presence of armed guards, properly trained, together with all other precautions to delay the entry and exit of would-be robbers probably will deter attempts. In such a case the weapons are justified, but only during the period when the risk exists.

Another situation involves the single guard in nighttime patrol in exterior areas or in high-crime locations. The possibility that the guard will meet a trespasser intent on some crime at or within the facility will expose that guard to a risk of his own life unless he is adequately protected. Again, the use of a deadly weapon would be justified.

Finally, there are those cases in which industrial guards must or should act as police or peace officers. What those cases are will be discussed next. To the degree that there are any such cases, the guard will be expected to prevent crime, apprehend criminals, and take all other police action required. The exposure level in that situation also justifies the issue of a deadly weapon.

Use of Deadly Force

If the industrial guard is within one of the exception situations noted, he will or may properly be issued a deadly weapon consistently, of course, with all appropriate statutory and regulatory requirements. What should the policy be in respect to the use of such weapons once issued? The legally permissible use of deadly weapons varies with the jurisdiction. Generally, however, all jurisdictions recognize the right to use deadly force in resisting actual or reasonably anticipated threats of deadly force against the guard or one whom the guard has a duty to protect. *That should be the only basis for use of deadly force by industrial guards.* Some jurisdictions allow the prevention of certain felonies or the prevention of the escape of a felon as grounds for deadly force. However, the current direction of thought is that use of deadly force for any reason other than to prevent its use by another is more than restraint or prevention of crime and partakes of punishment or retribution, both of which are essentially dependent upon prior judicial determination of guilt. Defense of industrial property or prevention of a crime against that property or capturing or recapturing a person who committed such a crime does not offer sufficient reason for the use of deadly force and the presumptive possible death of a human being. Even if the law of the jurisdiction would permit some extension of the limited application, sound industrial management will avoid the intentional taking of life.

Weapons and Training

If, in exceptional situations, deadly weapons are issued to industrial guards, adequate training in the care and use of the weapons, including clear communication of the policy in regard to that use, must be provided by the enterprise. Use of a deadly weapon by an industrial guard with resulting death or injuries, coupled with a showing that the guard was not properly trained, could result in civil and criminal actions against both the guard and the employer. If it were found that he had been wantonly or grossly negligent in his failure to provide or assure such training, the employer might be denied recovery of damages assessed against him even though he had adequate insurance coverage. The reason for the denial would be the public policy consideration forbidding one to be indemnified against his own gross negligence. It is not inconceivable that civil damages might amount to $1 million dollars or more in certain fact situations.

Even if the guard is a contract guard, there is the possibility of the guard's gross negligence being attributed both to the agency and to the firm utilizing the guard. Modern jurisprudence is more and more finding ways to support actions against all parties to any injury who could or should have acted to prevent the injury. It is important that the enterprise manager assure himself that the guard agency, if it provides deadly weapons to the assigned guards, is also providing training. A warranty by the agency that it does provide such training, inclusion of the client enterprise as a named insured under the agency's liability coverage, and adequate liability coverage in the client enterprise's own insurance portfolio are the appropriate precautions.

Industrial Guards as Peace Officers

Another major question in the use of guards is whether to seek peace or police officer status for them. Five main arguments are advanced in favor of such status. First, a peace officer is generally allowed wider discretion than a private citizen in making arrests or apprehensions. Second, the deputation often carries with it the right to carry deadly weapons. Third, the status as peace officer may open channels of communication not otherwise available. Fourth, there is a psychological advantage to a guard being known as a peace or police officer. Fifth, certain activities such as the direction of traffic upon a public road are permissible only if the person acting is a police officer.

The arguments against peace officer status are these: First, the constitutional requirements of due process will operate upon industrial guards if they are also agents of government. Second, the guard personnel may be subject to emergency draft or mobilization orders of the local police commissioner or chief. Third, failure of the guard to act in the face or presence of crime may expose him to legal liability both criminal and civil if he possesses peace officer status.

Except when industrial guards are also the only police force available, as they may be in very remote installations, and except for the limited application of the stated advantage of traffic control, the case against peace officer status, at least for the average guard, seems stronger than the case for it. Moreover, there are other means of dealing with alleged benefits deriving only from peace officer status. For example, with regard to the wider arrest or apprehension discretion, there are laws in a number of states that permit industries that meet certain tests in relation to operations in support of public welfare or national defense to allow guards and/or supervisory personnel to make arrests or apprehensions without broad liability to civil or criminal liability. For example, in New York State, a provision of the Defense Emergency Act, a type of war powers act, permits any employer certified by the state commissioner of transportation to post premises occupied by him with signs reading "No Entry Without Permission by Order Under State Defense Emergency Act."

Once a no-entry sign has been posted, entry into the premises by any person subjects that person to detention by any guard or supervisor of the employer for purposes of determining the person's identity and reasons for being present. If the guard or supervisor has reason to believe from the answers or behavior of the detained person that he did not have permission to be present, the guard or supervisor may arrest such person without a warrant or may summon a police office who may arrest him without a warrant.[1] Such statutory provisions greatly enlarge the permissible arrest and apprehension powers of an industrial guard without the need for peace officer status. Not every state has similar statutory provisions, but a number of the major industrial states do have them. A search of local statutes should precede a final determination on the question of peace officer status for guards to extend or enlarge areas of guard activity in respect to arrest and apprehension.

In regard to arrests and apprehensions generally, all states allow private citizens to make arrests under certain circumstances.[3] Thus, when a felony is being committed or attempted in the presence of the citizen, he may arrest. He may also arrest when a felony has been committed, even if not in his presence, and he arrests the actual felon. Some states allow private citizen arrest for a misdemeanor committed or attempted in the citizen's presence. In each case the citizen acts at his peril. The crime must in fact have been committed or attempted and he must have arrested the criminal. A properly trained industrial guard force will recognize situations in which crimes are being attempted or committed and thus will be on safe ground in making arrests as private citizens. Some further protection will be available through special liability insurance that the employer (or guard agency) can purchase. If the situation is one in which there is doubt as to the propriety of a citizen arrest, it is better left to the local police anyhow.

The argument regarding weapons as a justification for peace officer status is not persuasive. If the limited grounds for deadly weapons are present, permits can be applied for under local law. An exception might be made in a state that would not permit or does not license anyone but peace officers to carry concealed weapons.

The problem of constitutional due process operating against guards possessing peace officer status is an important and relatively new consideration. Since the landmark decision in *Miranda* v. *Arizona*,[2] increasingly strict requirements have been imposed upon police with respect to the procedural aspects of criminal investigations, arrests, interrogation, search and seizure, and the admissibility of evidence upon trial. There are numerous pitfalls into which an unprepared police officer can stumble, from failure to warn a suspect of his constitutional rights against compulsory self-incrimination to the unauthorized extension of a search.

The constitutional principle is clear that procedural due process is required to protect the citizen against the acts of the government. But an industrial guard, when acting under a deputation as police or peace officer, is acting as an agent of the government. The fact that he acted primarily to protect industrial property or in a situation that arose only because of his function as an industrial guard will not relieve him of compliance with every procedural restraint imposed upon any police or peace officer. As a result, by improper conduct, a guard can prejudice a successful arrest and prosecution. Unless there is strong reason for the appointment as a peace officer, the wiser practice would be to forego such appointment.

A compromise is possible for all these questions if, as previously recommended, a strong supervisory echelon exists. The peace officer appointments and license to carry weapons may be considered for the supervisory group, for whom the additional training will be a sound investment. If a supervisor responds to a situation in which he is called upon to act as a peace officer, in most jurisdictions he can demand assistance of citizens in the vicinity. The guards can be those citizens, and the protection of the law can be achieved in large measure. Of course, the same cautions apply to the propriety of the supervisor's conduct. However, increased training, broader experience, and superior ability will sharply reduce the likelihood the supervisor will prejudice the situation through ineptness.

REFERENCES

1. Unconsolidated laws of New York, Secs. 9163, 9165.

2. *Miranda v. Arizona,* 384 U.S. 436.

3. Cf. generally *Scope, Legal Authority of Private Security Personnel,* report Private Security Advisory Council Law Enforcement Assistance Administration, U.S. Dept. of Justice, 8/76, especially appendixes C-1, C-2.

Self Test Questions

1. When is the use of guards appropriate as a security countermeasure?

2. How can you quickly establish the total guard personnel schedule in a facility, assuring that each guard assigned works standard shifts and hours?

3. How can scheduled meal periods be used to increase available personnel at peak times?

4. What are some typically cited advantages to a proprietary or employed guard force as distinguished from an agency or contract guard force?

5. What are some disadvantages to the employed force?

6. Identify criteria to use in determining whether to opt for an employed or contract guard force.

7. What five factors will substantially impact upon a guard supervisor's effectiveness?

8. What criteria should control the decision to issue lethal weapons to a guard force?

9. What are the chief arguments for and against seeking peace officer authority for private guards?

= 9 =

Bombs and Bomb Hoaxes

OF all the security problems facing a manager, the bomb is the most dramatic. The bomb is simultaneously a danger to life, property, and the capacity to continue operations. A bomb threat triggers the instant emotional reactions of fear and consternation.

United States history is liberally peppered with incidents of bomb violence despite the impression that bombing is a very recent phenomenon. A staff report of the National Commission on the Causes and Prevention of Violence[1] contains an extended discussion and review of various forms of violence in America beginning even before the American Revolution. The major difference between earlier and present bomb violence is the lack of boundaries to the current problem. Bomb threats and bombs (the latter a very small percentage of the former) have troubled every major United States city in the past several years and have been attributed to political, social, racial, ecological, and economic militants as well as to the inevitable psychopathic population. Any extreme group may resort to real or threatened bombs against any target in support of any position.

With such widespread adoption of bomb violence, every businessman or manager may face the problem at some time. Every enterprise should be prepared for that moment. Geographical location, popular image, and past history are no longer reliable measures of susceptibility to bomb violence. Even the remotest connection with a cause or grievance—sometimes nothing more than being the largest enterprise in the vicinity and the natural focus of public attention—renders an organization vulnerable.

A Basis for Planning

Because the bomb threat is so emotionally charged, relying totally upon improvisation to guide response will lead to bad decisions. The response framework must be established beforehand by using rational assumptions. However difficult an on-the-spot decision may be regarding shutting down or evacuating, it will be more difficult if the decision maker has no standards on which to rely. It is often argued that because any bomb threat might be the real thing, response must always assume it is. That argument is as often followed by another: reduce the risk to life, evacuate, at least temporarily. The second proposition is not a necessary corollary of the first, nor is it sound in many cases. The reason the two are coupled is that, without preplanning, most persons will initially react to a bomb threat without critically testing its credibility and without evaluating any intermediate response. The "safest" thing to do is to evacuate; therefore, evacuate! In fact, there are situations in which evacuation may be a very unsafe thing, even in the face of a credible threat.

Planning to deal with bomb violence should be a part of overall planning for disaster emergencies. Three reasons suggest this. First, the effort needed to develop a comprehensive bomb plan is considerable. But bombs are only a part of the disaster emergencies a given business may face, and expensive and difficult planning should cover as broad a set of problems as possible. Many countermeasures required for dealing with bombs will also be appropriate for other emergency situations. A carefully developed plan will include all threats. The second reason is the greater likelihood that a total plan will remain current and be subjected to tests and reviews on a regular basis. Standing alone, a "bomb plan" acquires an air of unreality. Until the actual threat occurs, it is not used and often not thought about. At the time of threat its prior disuse may result in confusion. Early planning for civil defense encountered the same difficulty. Later doctrine emphasized incorporating civil defense into regular emergency planning.

The third reason for generalized disaster planning is the enhanced continuity such planning gives to actual disaster response. A broad plan involving all logical elements of the organization in scaled response to progressively serious emergency situations will permit smooth transition from normal to emergency operations. The savings in human and material resources can be great.

Elements of the Plan

The most serious bomb situation will be one in which an explosion actually occurs, but the more likely situation will involve only the threat, often inherently incredible. Planning must cover completely the aspects of response that will occur in all cases. The final steps—what to do if a bomb is found—often command disproportionate and premature attention. Disposition of unexploded ordnance is a technical and highly dangerous activity in which the enterprise will play a minimum role. Those who will be active in even that minimum role will be a readily identifiable group for whom special provisions can be made in the plan. On the other hand, every employee and visitor on the premises may be involved in an evacuation. Planning emphasis must be upon the events and circumstances most likely to occur and involve the largest numbers of persons.

Receipt of a Bomb Threat

The first consideration should be of how and when a threat might be received. Usually bomb threats come in anonymous telephone calls. However, they could be mailed or even be surreptitiously hand-delivered. In each case the prime questions are these: (1) If a bomb threat were made, who would be most likely to receive it? (2) Are there times at which no one would be in a position to receive it—for example, at night in a closed office or plant? (3) What can be done to prepare those most likely to receive a threat to deal with it?

In businesses with a publicly listed telephone number and a switchboard operator or staff, a high probability exists that one of the operators would receive the telephone call. However, in establishments with extensive direct-inward-dialing telephone service, a caller might inadvertently or intentionally reach almost any telephone. In that case some preparation should be considered for everyone in the organization.

There are many firms that do not maintain night switchboard facilities and that do not operate at night. A caller would be unable to reach such a firm with a bomb threat. However, hundreds of bomb threats are delivered to police and fire departments but directed at business and other locations. It is important that the local police and fire departments have a means of emergency communication with the business. The plan should include emergency liaison and notification provisions. A comprehensive plan will also consider communication requirements for fire, flood, hurricane and all other relevant disaster considerations.

Public emergency service and law enforcement agencies will take appropriate action when they receive a bomb threat. The manager's concern is with the preparation of his own personnel. The checklist that follows covers the items most important in dealing with an anonymous threat. Telephone operators and others key communications personnel should be drilled in the checklist. Test exercises are an excellent training device if prudently managed and confined to the key groups. Other employees, less likely to receive the threat, should be prepared at least to the extent that they are familiarized with the list and have an opportunity for occasional reinforcement or review, as through a poster or notice.

BOMB THREAT CHECKLIST

As soon as it is clear the caller is making a bomb threat, LET HIM FINISH HIS MESSAGE WITHOUT INTERRUPTION! If any response is essential, as to a statement such as "This is about a bomb, are you listening?" keep it to one or two words. While the caller talks, get the message EXACTLY and also listen for clues to:

1. Caller's *sex* and *approximate* age
2. Noticeable condition affecting speech such as *drunkenness, laughter, anger, excitement, incoherency*
3. *Peculiarities* of speech such as *foreign accent, mispronunciations, speech impediment, tone and pitch* of voice
4. Background noises audible during the call such as *music, traffic, talking, machinery*

When the caller has given his message, try to keep him in conversation. The following are key questions and should be asked, if possible, AFTER THE CALLER HAS GIVEN HIS MESSAGE:

Where is the bomb located?
What time will it explode?
When was it placed?
Why was it placed?

Note whether the caller repeated his message or any part of it. Note the exact time of its receipt. Write the message down IMMEDIATELY after the call. IMMEDIATELY after that, notify [*name and telephone number of appropriate person*]. Repeat the message EXACTLY AS YOU RECEIVED IT; then fill in the other details you were able to get.

BE CALM! LISTEN CAREFULLY! REPORT EXACTLY!

There is some possibility (though not much in isolated threat cases) that a telephone call can be held long enough to establish the calling number through tracing. Tracing requires immediate action by telephone company personnel and more time than is usually available. For a trace to be successful at all, the plan must provide a way for the receiving personnel to signal others with immediate access to telephone communications facilities that a bomb threat is being made. The person (or device) receiving such a signal would then immediately connect to the telephone company on a different circuit, identify the circuit involved in the threat, and request trace assistance. A preplanned signal to another operator or supervising operator could be the trigger for that response. How the signal is communicated depends upon the kind of telephone equipment involved. This phase of the plan should be coordinated with the telephone company servicing the enterprise.

There may be a chance of successfully identifying the caller if a complete sound recording of the threat is available. Installation of a solid-state audio recorder in the communications network, to be activated upon receipt of a threat and selectively to record that circuit, would improve investigation prospects. Such installation should be made in a way completely compatible with federal and state laws and with tariff requirements affecting the telephone company. Again, consultation with the utility in the course of disaster emergency planning will cover the point.

Initial Action After Receipt of the Threat

After a threat has been received and screened, the next consideration is what to do with the information. There must be a responsible person always available to whom it can be given for action. Many plans founder here because they allow only unrealistically high level personnel to take the response action. If the plan calls for the president or executive vice-president to handle a bomb threat, what happens if neither is available? Would the switchboard operator have to try a series of alternates? Would she know the proper person to try next? Would any person she might try be prepared for the responsibility?

An emergency must be dealt with by persons present or immediately available. The best approach is to establish a list of persons authorized to deal with the problem in the order of preference for notification. Then, and as an absolutely necessary part of the scheme, a technique should be selected to tell at all times who on that list are available. A duty board or in-house notice is an effective way to do it. Some one person must be responsible for assuring that the board or notice is up to date. For every minute lost in transmitting the initial threat to a responsible person, management options for action are reduced.

When the message has been relayed to a responsible person, the next concern is to evaluate the threat. This is the single most important step in the bomb procedure. The key to whether an enterprise deals rationally with a threat or rushes into over-reaction depends upon the skill with which the threat is analyzed. But a sound plan may at times require analysis and related action decisions from a relatively lower ranked executive or supervisor. Because he will not be experienced with making decisions involving major costs or total response, he will need at least some well-drawn guides to help reach his decisions. He must also believe that, whatever his decision, if he follows the guides and uses his best judgment, the organization will support him.

Assumptions

Although the decisions will always depend upon the facts in the particular case, there are some general assumptions that will be helpful.

1. Actual explosive devices are more likely to be found in buildings that are completely or partly accessible to the public than in those that are not. Experience to date establishes this probability.

2. The more effective the controls on admittance of persons and materials, the less likely it is that an actual explosive device will be found in a given facility.

3. The favored places in which explosive devices will be placed are those in which a member of the public might expect to find privacy or shelter while placing the device. Lavatories, utility closets, electrical and mechanical lockers and elevator shafts and pulley houses have been used.

4. If a device is planted and does explode, damage is more likely to be confined to a specific area than to involve the entire facility. That is especially true in very large facilities and in those with good fire dividers or strong bearing walls (Figure 10).

5. Dramatic response to a warning or threat, as by evacuating all or part of a building, is likely to be followed by additional threats.

6. The more detailed and credible the warning the greater

Figure 10. Damage to a Room Caused by a Bomb Consisting of Three Sticks of Dynamite

the likelihood that an actual explosive device is involved. If a warning is given in connection with a genuine bomb, it is because the bomber is intent on avoiding or reducing injuries to persons. In such a case he will make a positive effort to be credible and will provide enough information to convince those to whom he speaks. One intending death or injury will not warn.

7. A nonspecific or inherently incredible warning (for example, from a child, a drunk, a giggler, or an incoherent person) is not likely to involve an actual explosive.

8. No police or other public agency should be expected to provide any significant help between the time a warning is received and the time an actual device is found or an explosion occurs. Most enterprises will be dependent on their own resources, both as to decisions to be made and post-warning precautions to be taken. The police and fire organizations must be notified, however.

9. Because of possibilities for disguise and concealment, the only persons likely to locate an actual explosive device through a search are those who are intimately familiar with the area being searched. The only reliable criterion in conducting a search is that any object capable of being or concealing an explosive (generally anything the size of a typical lunch box thermos bottle or larger) is suspect until identified by a person who recognizes it. Figure 11 illustrates the shape and size of explosive devices recently used and indicates the relative ease with which they could be concealed.

Figure 11. An Actual Unexploded Pipe Bomb

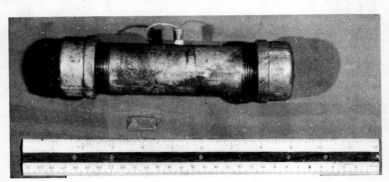

Evacuation

The most frequent question in bomb scare discussions is, "When do you evacuate?" There is no single answer. One answer might be to evacuate in every case. That answer, although it appears to assure the greatest protection for life and person, is not adequate. In a large enterprise, evacuation may bring hundreds of persons through a door or past a closet, logical bomb locations, where an actual explosion would cause more, not fewer, injuries. It is also pointless to attempt to evacuate a multistory office building in a metropolitan city housing thousands of persons if a credible threat gives 10 or 15 minutes' time. Major office buildings can take an hour or more to empty. Because the "always evacuate" policy does not necessarily assure better personnel protection, the very considerable economic and practical impact of invoking it should be weighed quite carefully.

Another answer might be, "Never evacuate." That answer is even more unsatisfactory than the first one because it leaves no room for dealing selectively with the highly credible, localized, and specific threat affecting large numbers of persons. Moreover, there is a major question of civil and even criminal liability to an organization that categorically refuses to evacuate despite circumstances.

The reasonable answer in each case depends upon three sets of factors: (1) threat warning factors, (2) occupancy factors, and (3) time factors. A balanced analysis of all the factors should indicate whether an evacuation is required and how extensive it should be.

Threat Warning Factors

Inherent credibility of the threat is of prime importance. Is the threat specific about either the location or time of explosion? If it is specific as to location, could the involved area be partially evacuated without disrupting the whole organization? If it is specific as to time, does enough time in which to organize and conduct any evacuation remain? If it is specific as to location, is the named location one to which a bomber could reasonably have had access? Does the location really exist? To give credibility to a threat, a hoaxer may attempt to sound specific but reveal his lack of knowledge of internal arrangements and locations. If the threat is specific as to location, would a search by informed persons probably find the device? Could such a search be made immediately?

Nonspecific warnings or threats do not reasonably require evacuation. The larger or more dispersed the enterprise the less justification for evacuation. Evacuating large numbers of persons in vertical or horizontal patterns (depending upon whether the building involved is multistory or low profile) will increase population density in halls, stairwells, exit corridors, and other common traffic aisles. If the explosive is in one of those places, exposure will be increased by the evacuation. In such cases the evacuation route should be modified or controlled to avoid likely danger spots. But with nonspecific threats that is not possible.

Sometimes threats are received in series. That is common during labor disturbances and has also been noted in connection with employee terminations. Malicious or disturbed persons will continue threats as long as an evident response occurs. Each evacuation will prompt the next threat. At some point most businesses stop evacuating on the assumption that the latest in a series of hoaxes is more probably another hoax than a live bomb. The problem with that approach is the difficulty of establishing probability accurately. The first threat must be evaluated without comparisons. Subsequent threats should be similarly evaluated. If a threat would prompt evacuation based upon its own content, then the mere fact that is one of several should not cause it to be disregarded.

A rapid succession of threats—three or four on the same day, for example—would indicate a lower level of probability that the second and subsequent were genuine, particularly if evacuation or other obvious response followed the earlier threats. Some major firms and large buildings are receiving multiple threats every day and as many as a dozen after actual bomb violence or widespread publicity of a threat. When the threats occur over longer time periods, a week or more, there is less basis for assuming they are related. Evaluate each threat on its own content first; then weigh the effect of prior threats. Note, too, that the decision being discussed here is only that about evacuation. Other actions, particularly search action, will be taken even in cases of highly improbable threats.

Occupancy Factors

If the threatened premises are an explosives plant, greater caution is demanded than if a light-hazard company were involved. If the location is small, a different response may be appropriate than if it encompasses hundreds of thousands of square feet. The prime factors relating to occupancy are (1) size, (2) inherent hazards such as explosives, infectious organisms, or noxious gases, (3) the number of persons present, (4) the available communications resources, (5) the available emergency service personnel.

A characteristic of recent genuine bomb threats has been the choice of time of limited human occupancy for detonation. Property damage and notoriety have been the bombers' objectives, not injury or death. If a bomb threat mentions a time when few people are present, evacuation should be easier. In fact, evacuation then could be most appropriate, not only because of the coincidences of genuine bombs and periods of scant occupancy, but because the factors making evacuation a great problem would not be present.

There are occupancies that simply do not permit evacuation without as much risk of injury or death as might be produced by a genuine bomb. A surgical operation in progress does not permit evacuation of the hospital operating suite. A steel mill in the middle of tapping a blast furnace can't just stop. Those cases require a different basic approach to disaster emergencies. If evacuation is not a feasible option in case of bomb threat, then physical and procedural security measures must be adequate to assure there will not be a bomb. Site hardening, personnel access control, package and materials control, and frequent inspection or constant surveillance monitoring are among the countermeasures available.

Reassembly and reentry of an evacuated site are necessary considerations in the emergency plan dealing with bomb threats. If a large building in a metropolitan center is evacuated, how can the occupants be notified that return is safe? Evacuation requires more than merely emptying a structure. Assembly sites and rally points at which persons can group for instructions are needed. Means of communicating with rally points are needed; they can be human messengers, telephones, radios, or other remote means of communication. Specific persons both at the rally point and within the structure must be assigned responsibility for maintaining communications.

Partial evacuation is a more common recourse than total evacuation, particularly in multistory, high-rise locations or in industrial sites covering large acreage. A general rule has been followed by municipal police and property managers that if an evacuation is deemed necessary for any floor of a high-rise building, then the floors immediately above and below must be included. For a localized threat to low-profile structures, an entire area within structural fire walls or an entire building in a complex would be appropriate for evacuation.

The evacuation route must be preplanned too. Would persons attempt to walk down from the fiftieth floor of a skyscraper? All fire codes exclude elevators as means of emergency egress from burning structures. Most fire departments advise against the use of elevators, even if operative, during a fire emergency. However, for purposes of evacuation before a fire or explosion, they may be the only practicable resouce. Planning must include a means to override normal demand and make the cars available for threatened floors. This requires coordinated activity by the building manager and the tenant occupants.

The extent to which all these occupancy factors have been considered will help determine the feasibility of an evacuation. Evacuation may not be feasible at all, or it may be feasible under some circumstances but only with extensive advance arrangements. The person responsible for making the evacuation decision should know the occupancy constraints and the available, planned options.

Time Factors

If the threat is, "A bomb will explode in the computer center in five minutes," and the computer center is on the tenth floor of a building in which the affected firm occupies only some floors, what response is appropriate?

How quickly can the warning be transmitted to the computer center? How quickly to adjacent areas? How quickly to adjacent floors, particularly if not occupied by the threatened firm? What has the building management planned for such an event? How quickly could the center be evacuated after receiving the word?

There may be enough time to empty the computer center but not the whole floor. Or the threatened firm may be able to reach all of its own people but not the other firms. Here, clearly, the fact and extent of any evacuation is related to available time. A workable rule of thumb is that a credible threat, with little warning time, should require first priority to the immediately threatened area. After that, other locations can be warned in order of proximity to the center of danger. No warning action that would result in persons remote from the threat preempting escape routes or resources should be taken.

If a threatened explosion does not occur, how much time should elapse before reentry can be considered safe? Again, there is no certain answer. Delay of explosion may be due to error in setting time devices or to their malfunction. It may also indicate that the threat is a hoax. It is important not only to consider how much time has elapsed but what has been done during that time.

If a threat specifies a location, even generally, such as a floor, or a department, and a time of explosion, then some options may be available. If there is enough time until the announced detonation, a search of the affected area may be made. If the search does not locate an explosive or suspect device but evacuation is still ordered because the search could not be regarded as conclusive, reentry should be deferred long enough to allow for bomb-timing malfunction or error. Given a midmorning evacuation with an explosion threatened for 11:00 A.M., general reentry after the usual lunch period, at 1:00 P.M. or later, would be reasonable, provided that during the period of evacuation the suspect area was controlled against any entry and provided the search was thorough.

If there is not time to make a preliminary search and the evacuation is considered necessary, then reentry should not be considered in any event before a thorough search has been made after the threatened time of explosion. The search itself might reasonably be deferred for a period for the safety of the search team.

If the threat is credible enough to warrant an evacuation, premature reentry to save productive minutes is irresponsible. The key decision is the one on evacuation. A decision in favor of leaving should not be neutralized by one to make premature reentry. The aftermath of a bomb scare requiring evacuation will involve extended conversations among persons evacuated, speculation about the bomb, and general lack of attention to usual tasks. It is illusory to assume that productive activity will be resumed promptly just because people return.

Optional Evacuation

An increasingly general practice is for the threatened enterprise to communicate certain warnings to its work force and give each employee the option of evacuating as a matter of personal choice. For organizations in which it is possible, there is much to recommend that course. First, it does share a threat warning with persons possibly in danger. A policy of concealing bomb warning from the work force while at the same time conducting any kind of obvious search is unwise because the search activity will inevitably provoke questions. It may not be possible to conceal the threat. Having it broadcast without any evaluation or positive reaction by the firm invites panic and certainly builds ill will.

But there are situations in which optional evacuation is not workable. In process industries or assembly operations in which the work of one depends upon the work of another in a continuous chain, a break in the chain may be more serious than a temporary stoppage. In those cases all the work should stop. That means that evacuation will have to be announced. It re-introduces the whole threat analysis problem.

Improbable Threats

If a threat is credible, it will require some response. The response may be a limited or general search, a limited or general evacuation, or some combination. Search tactics will be discussed in more detail in a following section. However, because of their relationship to evacuation and because of the kind of curiosity even limited searches can provoke, they must also be considered in the context of evacuations and communication of warning.

A nonspecific bomb threat that offers no rational basis for either search or evacuation should not be communicated in situations in which optional evacuation cannot be considered. To broadcast the threat and not suggest or permit any response would heighten anxiety and generally result in efficiency problems without adding to the safety of the work force.

A threat that requires a limited search or one conducted without disturbing large numbers of workers should not be communicated until a post-search decision is reached. That decision may be that the search is inconclusive and that evacuation is in order. The evacuation signal will then constitute the communication. If the search is conclusive and no danger is probable, no general communication is necessary.

A threat that requires large numbers of workers to participate in a search or observe it in process requires a communication. If a broad search cannot be conducted without arousing great curiosity and if explanation of the search without simultaneous evacuation would probably produce fright or uncoordinated actions, the communication should be an evacuation signal. However, a search may be broad enough to require evacuation from one part of the plant without requiring total evacuation. In that case the partial evacuation should be carefully supervised to assure its limited scope and also to permit selection of all the evacuation routes at the time of the signal rather than rely on prearranged routes. This last precaution is necessary to assure open access for fire or other emergency service equipment to reach the site of the suspected bomb without disturbing other work areas not ordered evacuated.

Conducting the Bomb Search

A bomb of high energy potential can be a relatively small thing. Searches must consider even small items. In most enterprises there are many items small enough to be or contain a bomb. The only reliable rule for eliminating those that certainly are not or do not is for a person familiar with the area and its usual contents to search it. That is the chief reason why police and local municipal emergency service personnel are of limited help in the early stages of bomb threat response. But because police will have been notified in the very beginning, they may respond immediately and offer to assist in the search. An organization with a trained, professionally competent security staff may prefer that the police remain on standby while the firm itself completes the search. Two reasons for this position are the greater speed with which the security personnel, who are already familiar with the site, could work and the desire not to add to concern by the presence of uniformed police. When the police force has no special expertise in bomb handling, that is an even more cogent position.

Small firms, without security staff, may find the police vital. The early planning should cover this, and the firm's position should be made clear to the police when the plan is discussed with them. Whoever it is who conducts the search concentrates on the area identified in the threat, if there is one. The search technique is basically a visual inspection of an assigned work area to be performed by a responsible employee intimately familiar with it. Often that is a first-line supervisor. The search is limited to inspection of the area and mentally accounting for all items observed. A strange item that cannot be accounted for after limited, preliminary check must be regarded as suspect.

At that point, the search is turned over to competent explosives technicians either from the police department or from the nearest military ordnance organization (Figure 12). The police will identify the military resource available in the area during planning conferences. If the police are not trained or equipped for bomb disposal, the military specialists may be needed.

If a suspect item should be located, an evacuation of the affected area is accomplished immediately if it has not already taken place. The evacuation need not be to the exterior, but it must be to a safe distance. What a safe distance is depends in part on the size and location of the bomb and the nature of the area containing it. The explosives personnel will advise in each case. General guides are available from government-approved and government-assisted publications. One publication, issued by the General Services Administration, the federal real estate management agency, suggests the following: [2]

Explosive Weight, in Pounds	Evacuation Distance, in Yards	Dangerous Debris, Distance, in Feet
10	150	150
50	150	150
100	250	200

Figure 12. A Booby-Trapped Pipe Bomb Equipped with a Fake Fuse

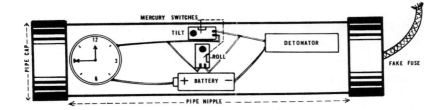

The bomb will appear to be a dud, but the clock on the interior can be so set that, when it reaches a preset time, the bomb is armed. One of the two mercury switches that are the triggering devices will then detonate the bomb with the slightest movement. One of the switches will detonate the bomb if it is tilted and the other if it is rolled.

Another publication by the National Association of Manufacturers, produced in cooperation with the Provost Marshal General of the Army,[3] suggests a minimum clear zone of 300 feet. Both guides can be modified for the presence of solid masonry partitions and other structural or natural barriers that offer a shelter advantage. Any doubt about whether an interior movement will provide adequate safety should be resolved by evacuation to the outside.

Some areas of some locations will always be suspect places, even in nonspecific threats; the toilets, lavatories, and janitor closets of most public buildings are among them. Maintenance, building service, or security personnel can be assigned to routine search of those areas while specific-area searches are made in other possible target locations. Locking such spaces and utilizing surveillance or intrusion alarms and devices whenever possible on a regular basis will reduce or eliminate the searches.

Some Legal Considerations

The enterprise that receives a bomb threat, at that very moment, has a variety of legal relationships with persons on the premises. Some of those persons are employees; others are business visitors; still others may be invitees of some kind. There may also be trespassers or persons attempting or committing some crime. The bomber would be one of the latter.

Response to the threat should be mindful of those relationships. The common law has always required some degree of care by a property owner or manager toward those on the property or likely to come on the property. What is that duty with respect to bomb threats?

Simply stated, in its most generally applicable scope, the duty is to do what a reasonably prudent man would do, in the circumstances, to prevent or lessen damage and injury. Of course, there is no model of a reasonably prudent man, so the measure of what should be done, ultimately, is what a jury or court, with all the benefits of hindsight and calm recollection, would find that a reasonably prudent man ought to have done. That is not so difficult a guide to apply, even though it lacks certainty. A reasonably prudent man would not neglect to make a plan for dealing with bomb threats in an era when such threats are fairly common. He would not neglect to consider how people could safely evacuate a building that was threatened. In short, he would not neglect to do those things that have been recommended as parts of disaster emergency planning.

In a given situation, determining the prudence of a decision whether or not to evacuate or search should be possible in the light of how well the decision agreed with the plan. A careful plan, reasonably drawn to include probable contingencies, must permit some discretion. The difficulties of inflexible plans and policies have already been mentioned. If an on-the-spot decision is shown later to have been wrong, it will not necessarily be found to have been unreasonable. It is not the final facts that determine reasonableness, but the facts as they would have been perceived by a reasonably prudent man at the time of the decision.

The development of the plan for disaster emergencies should command the best talent that the enterprise has or can obtain. Coordination and cooperation with local police and fire, other businesses, trade associations, and other persons with special competence are all highly important. Periodic rehearsal and test review of the plans are also important. Preparing the persons who will make key decisions by including them in the planning and providing them with as many reliable guides and standards as possible is fundamental. Then, when the plan must be followed in a genuine emergency, bomb threat or otherwise, the decision makers must be allowed the discretion required by the facts.

Insurance and Liability

The common-sense approach suggested in this chapter should be accompanied by early and continuing cooperation with liability and other casualty insurance carriers. The disaster plan and bomb threat procedure should be reviewed with them because they will share or assume total financial responsibility for damage and injury. The loss prevention departments of the casualty carriers will also be able to provide suggestions, particularly to the smaller firms without staff security resources. Review of existing coverage should also be made with an eye to possible damage, loss of use, and business interruptions that might be caused by an actual bomb.

Summary

The outline that follows was deliberately not set out at the beginning of this chapter because there are no short answers to the question of what to do about bomb threats. A checklist can, at best, only refresh the memory. The facts in each case will require careful analysis. All the standards must be considered and applied.

Before Receipt of a Bomb Threat

1. Develop an overall disaster emergency plan.
2. Include in the plan a priority list of persons to be notified in emergencies.
3. Be sure at least one person of those on the list is always available.
4. Consult the local police. Arrange for emergency notification from them if they receive threat warnings directed at you.
5. Analyze all your operations when planning. Identify any that could not be stopped or interrupted without serious risk of harm or damage. Consult your insurance carriers.
6. Provide physical security controls, at least for the operations identified in step 5, to assure that only authorized persons have access and all entering packages and materials are cleared.
7. Study the structure of the building or site. Identify areas that are self-contained and protected by solid walls or barriers. Plan evacuation routes and assembly points.

8. Assure communications capability for emergency warning or evacuation instructions to all personnel. Public address, telephone crash alert, bull horn, or human messengers can all be used as needed. Some standard evacuation signal for all types of emergency is better than separate bomb warnings.

9. Assign search responsibility to key personnel in all areas. "Search in place" is the proper technique. The task should be treated as a voluntary effort, and first-line supervision is generally the best source of personnel.

10. Prepare those likely to receive the bomb threat warning. Train them in the proper way to handle the call. Consider sound recording equipment. Consult with the servicing telephone company.

When a Bomb Threat Is Received

11. Get the message EXACTLY. After the caller delivers the message, try to learn WHERE the bomb is, WHEN it will go off, WHEN it was placed, WHY it was placed.

12. Get the message IMMEDIATELY to the available person responsible for handling it.

13. Notify the local police.

14. Evaluate the threat. Order evacuation if the threat is specific and credible and there is no time to complete a conclusive search. Don't evacuate for nonspecific or inherently incredible threats.

15. Search the target area. In nonspecific threats directed at public or semipublic buildings, search the halls, lavatories, unlocked utility and service lockers, stairwells, and elevators.

16. If evacuation is ordered, keep the evacuated area empty until the search is completed.

17. If evacuation is ordered, account for evacuated persons to the extent possible. Maintain some communication with waiting workers.

18. If evacuation is ordered, do everything possible to assure that self-help resources will be in place if needed. This

includes company medical personnel, building service and maintenance people, and security staff.

19. Do not return too quickly after evacuation. Allow a margin of error for malfunctioning or improperly set time delay devices.

If a Suspect Device Is Found

20. Be guided by the police or military experts. Do NOT attempt to move or neutralize the device.

21. If not already done, evacuate the area. If partial evacuation is appropriate, guide and control it to assure that routes of access for emergency service resources are not blocked.

22. Do everything possible to minimize possible damage from explosion. Shut down nearby operating machinery. Remove sources of combustion and combustibles, especially explosive and highly flammable materials. Marshal as much fire extinguishment first aid equipment as is readily available. If there is a plant brigade, have it stand by at a safe distance. Do everything your plan calls for to prepare for a major disaster emergency.

If There Is an Explosion

23. Upon arrival of municipal fire department follow directions of the senior officer. Until then, account for all personnel in the area and apply fire first aid to the maximum practical extent.

24. Have company medical personnel supervise removal and first aid treatment of the injured. Call public emergency medical help, particularly ambulances.

25. Arrange for notification to families of injured personnel. Location and last known condition are the most important items.

26. Have building service and maintenance personnel begin salvage and damage control operations as soon as it is safe to do so.

27. Notify insurance carriers.
28. Provide instructions to uninjured personnel about when and where to return.

After Each Bomb Threat

29. Review the plan. Review the response. Check for omissions in the planning. Revise as required.

REFERENCES

1. Report to the National Commission on the Causes and Prevention of Violence, *Violence in America, Historical and Comparative Perspectives,* Vols. 1 and 2. (Washington, D.C.: Government Printing Office, 1969).

2. U.S. General Services Administration order 3ADM 59302, December 31, 1963.

3. *Bomb Threats to Industry,* National Association of Manufacturers, undated pamphlet.

Self Test Questions

1. What reasons should you give for planning bomb responses as part of overall disaster emergency operations?

2. What would you consider when reviewing how a bomb threat might be received?

3. Describe nine assumptions which can be helpful in assessing any bomb threat.

4. On what factors does a properly made evacuation decision depend?

═══ 10 ═══

The Systems Approach
to Security

THE systems approach to assets protection is an orderly and rational approach to the planning and implementation of protection programs. It is being applied in increasing proportion in many industrial and business organizations. Two important benefits will usually result: cost reduction and improvement in the level of protection. The objective of the systems approach is to obtain a comprehensive solution to a total problem. The systems approach to problem solving was developed and refined through implementation of many U.S. Department of Defense and NASA programs. Large ballistic missile developments as well as the lunar and interplanetary space probes are examples of programs that have utilized systems techniques.

There are three general elements in the systems approach to security as follows: (1) a vulnerability analysis, (2) the selection of countermeasures and (3) a system test. The first step in applying this approach is to conduct a vulnerability analysis. That aspect will not be discussed in this chapter because the techniques of risk appraisal were fully developed in Chapter 1. Also, the system test will not be discussed as that aspect was considered at the end of Chapter 3.

200

After a risk analysis has been completed, optimum countermeasures that will perform the most effective job of prevention and detection within defined acceptable loss limits for the least expenditure of money should be selected.

Those who have designed security programs for business and industrial organizations in the past have often utilized the reverse process. Packaged solutions such as guard forces, standard design sensing and alarm devices have often been adopted as solutions to protection problems without prior consideration to the risks to be avoided. When the systems approach is utilized, solutions are considered only to the extent that they are specifically responsive to a demonstrated requirement. Only the elements best qualified to perform the assigned task most effectively at the least cost are designed into the system.

Because traditional methods of designing protection programs are no longer adequate, the disciplines and tools of systems engineering, advanced electronic techniques, and computer technology must be adopted more fully. Management in progressive organizations is beginning to realize that protection techniques that are antiquated by today's space-age standards no long offer adequate protection and are not cost effective. For example, it is not possible to cope with protection problems today with the single solution of the past—guards. It is prohibitive in terms of cost in today's labor market to attempt to provide an acceptable level of protection by simply increasing the number of guards on patrol and on post. Also, labor intensive protection programs are always the most expensive.

Even with the systems approach, though, it will never be possible to eliminate manpower entirely because it will always be necessary in any system to have personnel such as guards to provide intellect and judgment. However, because of the high cost of manpower, every effort should be made in the system development and design to eliminate personnel so that there will be an overall cost reduction in the protection program.

There was a constant increase in labor costs during the 1960s and 1970s and there is no indication that there will be any change in the future. Since the cost of manpower can become very sizable, management in every organization should be motivated to strive for maximum economy in the protection area through the utilization of labor-saving techniques and equipment. The systems approach can be the answer for most organizations.

Systems Development and Design

In selecting the best countermeasures that will be effective in avoiding risks at the least possible costs, three elements should be considered as follows: (1) the use of hardware, including electronics, (2) the utilization of manpower and, (3) the use of software. All three of these elements should be integrated into the design of the overall operating system.

Integration of Hardware, Software and People

Hardware components in a bewildering array are currently being marketed. Some manufacturers would lead the potential user to believe that each device or component being offered will be effective in the solution of almost all protection problems. Since a hardware item will usually give protection only against a particular hazard, each available device must be carefully examined to determine whether it is the best solution at least cost for the problem being considered. Also, it may be necessary to consider redundancy to assure complete protection against a hazard. For example, a combination of fire and security detection devices or several types of fire detectors might be necessary to give complete protection in a particular area of risk. In addition, a combination of other hardware devices such as locks, turnstiles, or fencing might be added.

After hardware components have been selected to protect against various hazards, the activities that personnel will assume in the prevention of the risks must also be examined. An effective response to any exceptions or violations detected must be considered.

Next, appropriate software to insure the effective operation of the system should be designed. The term "software" when used in this connection is intended to mean any written or verbal item needed. Examples of software items include policies, procedures, operating instructions, training, etc.

All of the three elements — hardware, manpower and software — will not always be required for protection against each hazard.

For instance, one electronic detector or a combination of such devices might offer adequate protection with little or no need for the interface of people. Mechanical devices, although perhaps less exotic than some electronic components, can be equally effective in giving protection and reducing manpower costs. A locked door, safe, or vault might be the only countermeasure necessary in a particular instance if properly integrated into a system design.

Electronic Protection Systems

There has been a general increase in the utilization of electronic components for security because the proper use of such hardware items will usually offer excellent protection in a cost effective way. A properly installed electronic component can be compared to the senses of a human and, with the possible exception of the olfactory sense, an electronic component is usually more reliable because it never sleeps and is alert at all times, wheras fatigue, emotional problems, or any number of factors might affect the alertness of an individual.

The electronic detector is ideally suited for many protection tasks because it is functioning as a device and is doing machine-type work. Consequently, it a mistake, both from the standpoint of efficiency and cost, to utilize personnel to perform tasks that can be handled otherwise.

There is a continuing evolution of technology in the electronics field. The state of the art has been constantly advancing at a very fast pace, and micro-electronics as well as integrated-circuit technology offers such benefits as increased performance, high reliability, long life, and low power consumption. The increasing digitization and miniaturization in the electronics field has also resulted in an increased application of computer technology and equipment in the business and industrial field. As a result, electronic components are continuing to become more reliable and the scope of their use is contantly expanding.

Because the use of electronic components has become such an important element in the design and implementation of security programs, the discussion to follow in the remainder of this chapter will be devoted to a discussion of the utilization of the systems approach in connenction with the use of such components.

The Technical Task — In the past, because those who dealt with electronic protection equipment usually utilized unsophisticated components and devices, electricians and mechanics could accomplish the design and installations. However, when modern techniques and equipment are being integrated into systems of any size or complexity, experienced, qualified systems engineers and designers should direct the development, design and installation. The individual components can, of course, be installed by mechanics and electricians under the direction of competent engineering personnel. Because of the number of different devices and subsystems that will usually be involved and the interconnecting networks that must be included, an electronic security system design is a complex technical problem. All elements must be integrated and related so that everything operates as a unit.

If the design is not considered a rather complex, time-consuming technical task, the results may be disappointing. The equipment, components, and subsystems might not operate well as an integrated unit with the result that the level of security might be less than expected and anticipated cost reductions might not be realized.

The technical problems can best be handled by a qualified technical organization that specializes in systems design and installation. The organization selected should have the capability of offering a turn-key system. Ideally, it should not manufacture or market equipment. If it does, its representatives will naturally be oriented to the equipment it is able to provide. If, on the other hand, it does not have a hardware bias, its technical staff members will be better able to choose the best item for each use at the most favorable cost. Therefore, it should be able to perform in the role of an architect-engineer. A better overall system should result.

Most companies will not have in-house capability to perform the technical work necessary to implement the systems approach. Organizations that do have a systems capability will usually have their technical staff members busy working on projects that will produce income for the company. In such a situation, the in-house technical staff would usually not be able to devote sufficient time to a security system development and design.

Since a system will ordinarily be designed to meet the requirements of a particular installation, an off-the-shelf system that will be effective usually cannot be purchased. Some manufacturers of protection equipment have used the term "system" to describe equipment designed to solve an individual problem. Such an application might technically qualify as a system because the equipment will offer a complete solution to a particular protection problem. However, when the term "system" is used as outlined in this discussion, an item or series of items designed to act as a countermeasure for an individual hazard might better be referred to as a "subsystem." The contribution to the overall protection program that can be made by such a device or component is often small.

Need for Technical-Security Liaison — The facility security staff representatives should work closely with the technical staff involved in the system development and design to insure that an adequate protection solution has been found for each system problem. As a matter of fact, there must be a close liaison between the two groups during the entire systems application that should end only when the turn-key system is accepted.

Security is a complex area of specialization, as has been discussed in preceding chapters. Although the systems organization may accept complete responsibility for an effective turn-key operating protection system, the technical personnel cannot be expected to become experts in security. If they do attempt to assume the role of security specialists, they will most certainly be operating in a vacuum. Without a constant dialogue between the technical staff and the facility security personnel, any system installation will almost surely not be as cost effective or as effective from a protective standpoint as it might be. On the other hand, the facility security staff members ought not attempt to become experts in the technical aspects of the system and attempt to perform or supervise any of the complex technical work.

An effective system can be realized only if there is a close exchange between the technical staff and the security staff with each performing in its particular area of specialization. Both the security staff and technical staff will, of course, gain considerable general knowledge about each other's area of specialization, but each group would do well to constantly keep in mind the old axiom that a little learning is a dangerous thing.

At the conclusion of the system design, the system technical personnel will be able to detail the cost of the system. Because of possible budget limitations or other problems, it may not be possible to install the entire system at one time. The advantage to this approach, if this type of decision is made, is that the facility will have a master protection plan and can implement it in any manner that seems appropriate. The implementation of the entire system might, for example, be planned on a milestone schedule so that the installation would be completed over a period of several years. By implementing a system in a planned manner, a piecemeal approach would not result and a hodge-podge installation would be avoided.

Some of the tasks involved in the close liaison that must exist between the technical and security staffs are shown in Table 7.

Systems Installation and Operation — The procurement of equipment and the installation of the system is the next task. The system technical personnel will be extremely busy at this point with technical problems that will include procurement and delivery of equipment, insuring that schedules are met, coordinating the equipment installation, and determining that the installation is being made properly.

Table 7. Systems Approach Tasks

Tasks	Technical	Security	Both
Make a requirements analysis			✓
Determine utilization of people and components			✓
Design the system			✓
Order equipment and insure delivery on schedule	✓		
Coordinate equipment installation	✓		
Develop operating procedures			✓
Program the system			✓
Train operating personnel			✓
Indoctrinate employees		✓	
Arrange for system maintenance	✓		
Debug the system	✓		
Operate the system		✓	
Plan the system expansion			✓

Because all of the tasks involved with the hardware installation are technical in nature, the facility security representatives will be of little or no assistance to the technical personnel at this point.

Software Aspects — At the same time the hardware is being installed, procedural components, or software, for the system must be developed and a training program must be designed and implemented for the personnel who will operate the system. The systems personnel must, in the overall system design, work closely with security personnel in the facility and assist the technical staff with those tasks. They can all offer valuable assistance and advice on training requirements and material needed as well as be effective in the development of software.

At this stage, planning the indoctrination of employees in the facility as one element of software is an important task that should be done. As was pointed out in Chapter 3 in the discussion of securiy education, employees must voluntarily accept all elements of the security program. It was also pointed out that the acceptance can be realized only if employees are knowledgeable and understand why particular controls are being imposed.

If all employees are not told how the new system being installed will operate, why it is being adopted, and what advantages will result for the employees as well as the company, some employees may not accept the new concepts because of lack of knowledge. Therefore, if an educational campaign is not planned as a part of the systems approach, the final systems operation could be adversely affected by the attitude of employees. If there were a lack of acceptance by a large enough group of employees, it is possible that even the best system would be doomed to failure. However, lack of employee acceptance should not be a reason for not adopting the systems approach, because the problem of acceptance is no more difficult to plan for than any other control or procedure that might be implemented to protect company assets.

Debugging the System — The final step in the systems approach is the complete operation of the system. Regardless of how well the requirements analysis and system development, design, and installation are done, it can be expected that some difficulties will develop when the system is placed in operation because of the complexities of the operating elements.

As a result, a period of debugging the system should be planned into the scheduling. During that period, all aspects of the system can be carefully checked to determine if all elements are operating as planned. At the same time, hardware and software mistakes in design can be corrected. Also, the controls that are being eliminated as a result of the new system operation should not be abandoned until the check-out period has ended and it has been determined that the new system is operating efficiently. As a result, for a period of time the old controls should be operated along with the new system. That factor should be considered in the planning of cost reductions, because any planned reduction of personnel will not occur as soon as the new system is installed; it will take place only when the new system is debugged and is in complete operation.

Expansion and Maintenance — After the system is in complete operation, need for improvements will undoubtedly become apparent. Also, the possibility of additions to the system will be recognized. As a result, it will generally be found that a protection program that has been developed through the application of the systems approach is a dynamic thing that will constantly be undergoing change. That, of course, is another benefit because changes and additions that become necessary can usually be accomplished quickly and at minimal cost once the overall system is operational.

Maintenance of the system should also be included in the system planning process. Since a protection system must operate around the clock, seven days a week, it is important that a rapid service response be provided. Replacement units and spare parts that can be expected to malfunction can be stocked in the facility. In the event of a component failure, the defective part can then be immediately replaced and sent to the manufacturer to be repaired or replaced.

The system technical staff should make arrangements for maintenance of the system as a part of the responsibility it assumes for the efficient system operation. It can do so by contracting with a reputable service company. Alternatively, the plant engineering and maintenance personnel in the facility can be trained by the technical systems staff to maintain the system.

An Application of the Systems Approach

The electronic systems approach has a particular cost effective application in the physical security area because of the possibility of

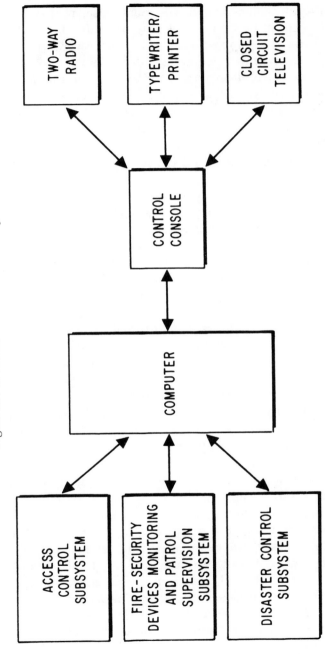

Figure 13. A Model Electronic Control System

significant manpower cost reductions. Also, a general improvement in physical protection can usually be realized. For the purpose of illustrating the application, a basic model control system that would offer protection in three areas will be discussed. The system comprises (1) control of personnel at entrances and other locations, (2) fire and security devices monitoring and the supervision of personnel who perform inspection and patrol functions, and (3) disaster control. The system is diagrammed in Figure 13.

The nerve or control center of the system would be designed to utilize a computer to control and supervise all of the system elements. The use of a computer in a physical protection system is now possible because of the rather recent development of small special-purpose minicomputers. Earlier computer applications involved expensive, large, central computer installations designed for multiple data processing application or for engineering and scientific use. However, because of reductions in recent years in the costs for minicomputers and peripheral equipment, it is now possible to consider the widespread use of such devices for economically feasible security applications.

The minicomputer is ideally suited to be utilized as the key to electronic information processing in a physical security system. Detection devices have traditionally depended upon the transmission of the information detected to a human for response and reaction. There was no other way to process such information until the development of the computer, which can perform logical functions at speeds many times greater than the speed of the thought processes of a human. A minicomputer can now process alarm signals or other information received at a control center in an automated fashion and also assume the responsibility for routine responses. As a result, an operator or supervisor at a control center can concentrate on nonroutine problems or the tasks that require intellect and judgment.

The systems approach principle discussed earlier in this chapter — assigning tasks to the best qualified resource — is followed when a digital computer is utilized to control input and output functions. A computer will not make errors in searching its memory for information; it can retain much more information in its memory than a human can; it can search its memory for a fact in a microsecond; and it can make logical decisions while performing many other tasks simultaneously. As a result, the minicomputer can electronically redirect and redeploy available resources to solve protection problems at lower cost, at greater speed, and with greater reliability.

The three elements in the model system will be outlined in the paragraphs to follow. Each of the three system elements will be referred to as a subsystem in the discussion. It should be emphasized that the model is being used only as an example of how the systems approach might be applied to the physical security area. It will not fit every organization that a reader might have in mind. Again, it should be emphasized that each system must be tailored to the particular requirement of each facility and that the implementing processes outlined in the earlier portion of this chapter must be followed, from the requirements analysis to the turn-key system operation, if successful results are to be expected. Thus this model is not being suggested as a package or off-the-shelf system but merely as an illustration.

Control of Personnel

The subsystem involving the control of personnel is often the key element in terms of cost reductions, especially in facilities in which security personnel have been utilized to control personnel and vehicular traffic at all entrances (Figure 14). The computer would be utilized to make decisions concerning the ingress to and the egress from the facility, and electronic and mechanical components would be combined at entrances and exits to interface with the computer to physically control those entering and leaving. The combination of the computer decision control and the physical controls at the entrances and exits would replace guard personnel deployed at fixed posts at the locations specified.

The personnel entrances and exits could be so designed that everyone entering or leaving would be processed through a double-door control booth (Figure 15). At least one door would always be locked. A card reader and a digit keyboard tied into the minicomputer would be installed in the interior of the booth, as well as a closed-circuit elevision camera that would be connected by cable to the control center.

An individual desiring to enter the facility would enter the booth, insert a machine-reader identification badge, into the card reader, and manually place into the 10-digit keyboard a four-digit number assigned only to that individual. The computer would be programmed to monitor each stage of the transaction and would lead the individual through each step of the entrace process by means of display lights while performing the necessary logic, timing, and control checks. If the combination of the individual's badge and the four-digit code met the programmed requirements for that particular individual in the computer memory, the computer would sequentially unlock the inner door of the booth, lock the outer booth door, and allow the individual to enter the facility.

Specified deviations from the normal sequence, of either events or time, would be signaled to the control center operator as an alarm condition. An individual exiting would be processed in a similar manner except that the keyed-in code might not be required. The booth would be under closed-circuit television observation from the control center, and a voice communication link between the booth and the control center would be provided. As a result, the control center

Figure 14. Elements Involved in the Model Access Control Subsystem

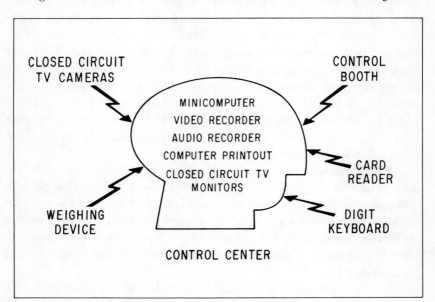

operator would at any time be able to observe any activity taking place in the booth, would be in voice communication with anyone in the booth, and would be able to override the automatic operation if necessary.

A number of special checks and logical sequences could be incorporated into the system for additonal security, for processing visitors, for monitoring property movements, or for use at especially sensitive locations or times. For example, a weighing device might be installed in the floor of the booth to weigh individuals as they inserted their cards in the reader. The weight of the individual, within perhaps a 20-pound range, would be programmed into the computer memory. As a result, an individual's actual weight would be automatically compared with the weight recorded in the computer memory. Although this added item would not be effective as a security control, it would be effective in alerting the control center operator if more than one individual attempted to enter through the booth at the same time. Or, it would give a signal if an individual were carrying a

Figure 15. An Example of a Double-Door Control Booth for Use at Entrances

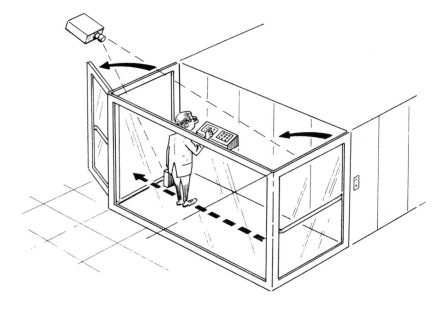

heavy piece of equipment, which might be of value as a property control mechanism in a particular application.

Because of the flexibility of computer applications, the controls mentioned here might be varied or changed to meet different criteria. Also, additional sophistication and added controls could be made possible through a slight variation in software at little or no cost.

The number of access booths would depend on the amount of traffic through each opening. Therefore, a count of traffic through each opening around the clock, seven days a week, would indicate the number of booths required at each location. The only limiting factor that would affect the operation of a booth would be the time required for an individual to perform all of the functions required and to go through the booth. The computer would not be a delaying factor, because it would be able to react without delay to each signal received.

Figure 16. Elements in the Model Fire Security Devices and Monitoring and Patrol Supervision Subsystem

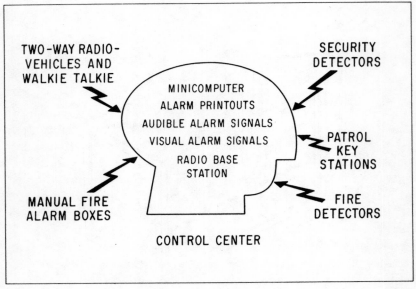

Device Monitoring and Supervision of Personnel

The second subsystem, fire and security device monitoring and the supervision of personnel performing inspections, would also be computer-controlled (Figure 16). Any exception, irregularity, or violation in this portion of the system would be signaled as an alarm at the control center by an audible signal, by a signal on a CRT, and by a printed record indicating the location and the time. The minicomputer would constantly monitor all security and fire detectors as well as all wire lines. Any wiring fault would be immediately signaled at the control center as an alarm. Some of the security functions that might be incorporated in this subsystem are:

- Door and window intrusion detectors
- Photoelectric and laser alarm detectors for perimeter protection
- Vibration, motion detection, or proximity devices for area security

Examples of some fire protection functions that might be integrated into this subsystem are:

- Fixed temperature, smoke, and rate-of-rise detectors
- Water flow and valve sensors
- Standpipe overflow detectors
- Water main pressure failure detector
- Carbon dioxide system sensors
- Foam or dry chemical system sensors
- Combustion sensors

The supervision of inspecting personnel would involve a variation of the "watchman supervisory service"— wall-mounted keyways installed throughout the facility that would be tied into the control center. As security or fire inspectors made their rounds on patrol or inspection, they would register their progress at each keyway on the route by turning a key in the wall-mounted lock switches. As the key was activated, it would signal the location of the individual to the computer. Since the patrol schedule would be programmed in the computer memory, the computer would check each time a keyway was activated to determine if the individual was on time or had activated it in an incorrect sequence. An alarm would be activated at the control center by the computer if any of the key reports did not match the program in the computer memory or if a keyway had not been activated in the time allotted.

A response to any violation or exception signaled at the control center would be planned into the system operation. Each individual on patrol would be equipped with a portable radio. Depending on the problem developing in the system, the control center operator could use the radio network to dispatch the number of inspectors or patrolmen necessary to handle the problem.

Disaster Control

The disaster control subsystem would be monitored by the computer at the control center and would include emergency evacuation speakers installed throughout the facility as well as the necessary controls at the center to assist with the handling of any emergency (Figure 17). The evacuation system would be controlled from the center by a series of switches so that the operator could alert personnel to an emergency in individual buildings or in all building on the site.

Figure 17. Elements in the Model Disaster Control Subsystem

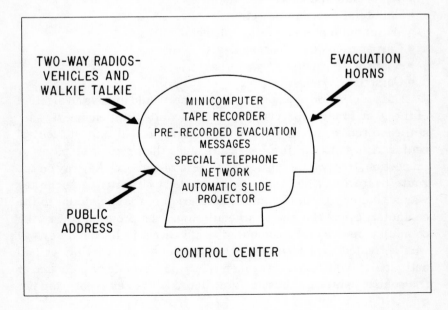

Tape recorders could be designed into the disaster control portion of the system. All emergency telephone messages coming into the center could be recorded. In addition, prerecorded messages that contained direction for evacuation which could be played over the evacuation speaker network might be included in the system design.

Special telephone networks could be included in the design. For example, direct telephone lines to key departments in the facility, such as medical and maintenance, might be considered. Special telephone lines could also be provided to outside agencies such as the police and fire departments.

Provision for the display of information might also be made a part of the subsystem design, as by the incorporation in the design of an automatic slide projector on CRT. Any kind of instruction needed by the control center operator could be programmed in the computer memory and be automatically displayed when needed. Such information might include emergency instructions, floor plans, telephone numbers, and the current location of key personnel. In addition, the computer could be programmed to perform any number of other functions automatically during an emergence. One obvious task might be to automatically dial key individuals. That would save time for the control center operators and allow them to perform less routine tasks.

Control Center

The control center would be designed to utilize the benefits of the minicomputer to the maximum extent. Information would be recorded automatically whenever possible. A video tape recorder would be utilized to record significant events on a selective basis or to record everything automatically. Selections of the mode would be computer-controlled. Audio tape recorders would be designed into the center to record all voice communications automatically. Alarm activities, access transactions, or other like system events would be logged on magnetic tape. A hard-copy printout of such activities would also be available.

Although the minicomputer would be designed into the control center console, the console operators need not be a computer operator. They would be trained to respond to system signals displayed at the control console and to manage that console only. Trained systems and maintenance personnel would have supervised access to the mini-computer when any software or hardware changes were needed.

The physical design and layout of the control center is important. The center should be as free of noise and outside distractions as possible, because those operating the system must be able to concentrate completely on information coming into the center. Operator fatigue and eyestrain must also be considered in the design of the center. Therefore, colors as well as lighting in the center must be given careful attention.

The display of information and the operator controls at the center must also be given careful consideration. For example, an operator should not be required to watch any system item on the console such as a television monitor constantly because of fatigue that could be expected to result from such concentration. Instead, if anything on the console requires the attention of the operator, the system design should provide for an audible signal to attract attention and a visual identification to indicate the source of the action.

System Expansion

As a system such as the model described became operational, it would usually be found that additional physical security tasks could be economically added. Also, it might be found that the system should be extended to other locations. That would be feasible because all of the signals in the model system, except closed-circuit television, could be multiplexed and transmitted on telephone lines. The only factor limiting the extension to other areas might be the cost of leased lines.

In addition to security applications, expansion of the system might also include such nonsecurity tasks as utilty monitoring, process control, personnel time recording, and project control (Figure 18). After the installation of the basic system, additional tasks can often be incorporated at a surprisingly reasonable cost.

Utilization of Manpower

As was pointed out earlier in this chapter in the discussion of the system development and design, the use of people must be considered along with the use of electronic and mechanical equipment in the systems approach. Therefore, the utilization of manpower must be considered to insure the effective operation of the model system.

Although one of the objectives of the systems approach is to reduce manpower costs, it is necessary that care be taken not to eliminate manpower to such an extent that the system would not operate effectively. Adequate personnel must be scheduled at the control center to insure that the exceptional situations or problems signaled at the center are properly handled by the personnel on duty there. Also, sufficient personnel must be available on patrol or assigned to other duties to respond and take corrective actions. The timeliness of such responses must be given serious consideration in the planning process.

Even after adequate manpower had been scheduled for the effi-

Figure 18. The Model Electronic Control System and Some Expansion Possibilities

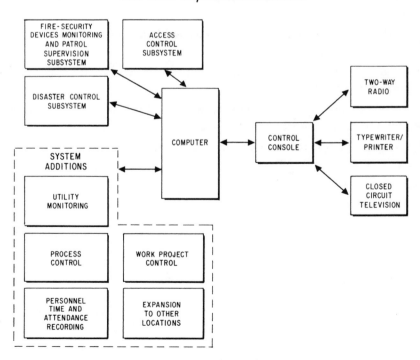

cient operation of a system, it would not be unusual for an overall manpower reduction of from 15 to 25 percent to result. Also, it would not be unusual to expect that the cost of a system would be paid out of savings realized from the reduction in manpower in two to five years.

Self Test Questions

1. What are two important benefits that will usually result when the systems approach to security is utilized?

2. After a risk analysis has been completed, what are the three types of countermeasures that should be considered?

3. Why is an electronic security detector, when properly installed, considered to be more reliable than a human?

4. Why should experienced, qualified systems engineers be utilized when a complex electronic security system is being designed and installed?

5. Why is the minicomputer ideally suited to process information in an electronic security system?

6. Name three areas that an electronic security system utilizing a minicomputer can control.

7. When an electronic security system is being utilized, list two types of tasks that must be done by an adequate number of personnel.

8. Why should expansion possibilities be planned into an electronic security system design?

=== 11 ===

Screening and Investigation
of Applicants

BECAUSE there is strong feeling about their ethical propriety and their practical usefulness, it is important to establish a convincing basis for utilizing applicant investigations. The recent public concern with invasion of privacy by government, business, and special-interest groups makes this subject more important now than ever before.

Reasons for a Screening Investigation

The chief reason for a background or screening investigation is to establish that the applicant has related all the relevant information and has related it truthfully. Assumptions of truthfulness about applicants concerning whom the prospective employer knows nothing except what the applicant himself tells him often result in serious difficulty when unfavorable information is later revealed. American political and general ethical convictions assume that each man is what he says he is. Regrettably, that is simply not true. A five-year study by a major employer involving over 6,000 applicants for employment showed that information was developed by investigation of applications that would not have been acquired otherwise and that contained major unfavorable aspects. Results of that survey were that of all types of applicants from shop hands to senior executives, 13.3 percent were shown by investigation to have some unfavorable background aspect that was not disclosed by the applicant. Of the 13.3 percent, 9.6 percent were seriously unfavorable to the point of rejection of the applicant.

Even more startling results appeared for applicants considered by class or category. Of all shop and office applicants, 15.5 percent had unfavorable background aspects, 11 percent seriously so; 9.6 percent of all scientific and technical applicants were questionable, 6.9 percent seriously so; and, most significantly, 18.9 percent of management applicants were characterized by unfavorable background data, 11.9 percent seriously so.[1]

In this single survey one in every ten applicants had such seriously unfavorable background aspects as to warrant rejection, and one in nine applicants for management positions required rejection. The information on which the rejections were based was not disclosed or suggested by the applicant, even though the opportunity was presented and even though the application forms and questionnaires were so designed that to omit or disguise the real data required a conscious effort to deceive.

Parallel data on employment applicants are furnished by the various commercial investigations agencies. For a number of reasons that will be discussed further on in this chapter, the results of applicant investigations that are conducted by mercantile or commercial investigating organizations often reflect a lower rate of unfavorable data than those conducted by appropriately qualified professional investigators employed directly by the prospective employer, as they were by the firm that published the survey data previously cited. The evidence from both the commercial agencies and the study cited indicates that there is a significant amount of information about prior employment that applicants will not disclose and that is relevant to the employment determination.

Excluding all information of a kind legally not permissible

as a basis for refusing employment, such as that prohibited by
the Federal Civil Rights Act and by local fair employment practices laws, prospective employers have the option of rejecting
applicants for employment at their own discretion. Except, then,
for reasons dealing with ethnic or racial origins, religious preference, age, or sex and except for the further requirements of the
National Labor Relations Act and similar state laws that there
be no discrimination based on the exercise of workers' rights
to organize and bargain collectively, an employer is unregulated
in respect to reasons for rejecting employment applicants.[5] But
there are many kinds of personal behavior and activities that
an employer might believe indicate unsuitability for employment.
It is precisely because of information related to such behavior
and activities that persons are rejected after the kinds of investigations discussed in this chapter. Because there are legally acceptable reasons for rejection and because the experience of some
employers substantiates their conviction that rejections are necessary, it is clear that applicant screening is useful.

Propriety of Applicant Investigation

As to the propriety of applicant investigations, it should be
noted that the employer, particularly the corporate employer
bound to conserve and make profitable the investment that sup-
ports the business, has an obligation to be prudent and reasonable in the selection of employees. But some employers consider
it unfair or an imposition upon applicants to conduct any investigation of the statements they make on, or omit from, application
forms. Reflection suggests that the contrary may more likely be
true. The applicant has had the opportunity to learn about the
company from all the sources of information available. In the
case of publicly owned companies, those sources are legion and
there is little about such a firm that an industrious applicant
could not learn before he made his application. The company,
on the other hand, has no information about an applicant until
he applies, and then it has only what the applicant provides.
To make a decision for or against employment on the basis of
such data alone would seem to be quite imprudent.

Technical evaluations and objective testing of one kind or another can help to establish the possession of specific skills and knowledge. But no testing will confirm statements about past history, nor, if the applicant is determined, will most interviewing that cannot go beyond the applicant's own statements. And it is not merely the fact that an employment applicant may be unsuitable for employment that should prompt a prospective employer to be certain before hiring. A variety of relationships arise concurrently with the employment that can threaten the employer's assets. For example, an employee is covered by workmen's compensation as of the date of hire. Short of self-inflicted injuries, the developing trend in compensation law is to permit awards for any and all injuries sustained at, in the course of, going to, or coming away from the business. An applicant's history in respect to unsafe practices and prior injuries is important.

Employers with health insurance programs beyond the workmen's compensation coverage are in even greater jeopardy from malingerers who have already demonstrated by past record an intention to take maximum advantage of such programs. Also, a narcotics addict can become a source of very serious difficulty by acting as a distributor within the business to support his own habit. In fact, that is rather characteristic behavior. Hiring persons with active histories of narcotics addiction can be a prelude to a widespread and continuing problem.

Another argument sometimes advanced against the use of applicant investigations is that they tend to make social rehabilitation impossible for persons with unfavorable histories by consistently eliminating them from consideration. But if the argument is that an employer should not protect himself against intentional deceit and has no right to information about applicants' histories, it upholds constant and unmeasured exposure to all manner of losses and so cannot be given a serious hearing. Such a position is not prudent. If, on the other hand, the argument is that the mere fact of unfavorable history should not automatically foreclose the applicant from consideration, it requires more appraisal. If social rehabilitation is to be possible, then those most in need of it will have personal or work histories that, at first blush, would render them unsuitable. However, many employers, including some of the largest United States firms, have agreed to assist in both the social development and rehabilitation of persons whose backgrounds would usually keep them out of jobs.

The very fact that an employer will be hiring persons different in important ways from the usual employee requires for the success of the effort of both employer and employee that all facts and circumstances affecting the relationship be evaluated fully and before difficulties that make retention of the employee difficult or impossible arise. It would be a disservice to a person who had been convicted of larceny, served a prison term for the crime, and had not since been employed to assign him to a position involving responsibility for or access to cash, not because the employee would necessarily steal again, but because, if after his employment a loss should occur, he would be a natural focus of suspicion. Yet it is not uncommon for firms to assign new employees from among groups with criminal or other anti-social histories to entry level jobs like messenger or mail room clerk. In most businesses a mail messenger has daily contact with and access to a variety of valuables in many places within the enterprise. It is one of the poorest job choices for persons with theft convictions. Comprehensive information about the nature of an applicant's record and careful consideration based on that information, including frank interviews with the applicant himself, would have a better chance of ultimate success than haphazard malassignments.

One other position unfavorable to applicant investigations should be examined. It is that employment, particularly of large groups of employees, can be likened to life insurance. Actuarial data will determine average life-spans. By certain kinds of personal habits and other exposures, individual persons will die from accidental or extraordinary causes much earlier than the tables predict. Notwithstanding that, life insurance companies, even those with only a casual applicant inspection policy, regularly profit from their underwriting. Similarly, an employer can average out his applicants by removing those whose later behavior indicates unsuitability.

The actuarial argument overlooks three critical points. First,

life expectancy can be predicted because of the vast number
of data readily available from vital statistics sources, whereas
career mortality cannot be because of the absence of the data.
Second, death is universal and happens once for everyone, but
discharge from employment does not. Third, the elapsed time
between hiring an unsuitable applicant and later discovery of
his unsuitability might be, and often is, sufficient for his conduct
to have cost the employer dearly before he is finally terminated.

For all the reasons discussed it is proper to consider employment applicant investigations a useful, prudent precaution, provided that such investigations comply with the law, are not the
basis for improper discriminatory practices, and are efficient
enough to produce a high degree of confidence that they will
serve their primary purpose.[6]

The Basis of Investigations

Some employers do not articulate their employment criteria
or standards of suitability, but they do make regular judgments,
through employment management or similar functions, about individual applicant suitability in particular cases. In some situations of this type, the employer even uses preemployment investigations as a screening tool. The problem with this approach is
that every decision is entirely subjective and the single discretion
of one or two persons determines the enterprise policy on hiring.

If an applicant is unsuitable for employment, it ought not
to be because the employment manager or some operations manager for whom the applicant would work thinks so. It ought
to be because, measured against some objective standard, the
applicant manifests a history or predisposition that makes his
employment a bad risk for that enterprise. The determination
of the employment or operations manager should really be only
that the behavior or predisposition is a fact and is present in
a serious enough way. Under those circumstances, the enterprise
policy is constant and the day-to-day decisions about its application can be reviewed in a rational way.

It is highly desirable to give major attention to the initial establishment of standards or criteria under which suitability will be judged. The final determination of what those criteria are should be a function of senior executive management, because they will reflect basic enterprise policy. The list below is a suggested standard and a set of criteria that have been used by employers. Readers who are familiar with the U.S. Department of Defense procedures for the granting of personnel clearances to classified information will recognize, somewhat rephrased, a number of the criteria that the government has used for years in making those determinations. The reason such criteria are appropriate for both industrial employment and government clearance is that they deal with basic aspects of personal suitability: integrity, sobriety, truthfulness, financial responsibility, and the like.

EMPLOYMENT SUITABILITY STANDARD AND CRITERIA

Standard

No applicant concerning whom there is serious unfavorable information in respect to any suitability criterion shall be employed except with express prior approval of executive management.

Criteria

The following suitability criteria shall be considered in evaluating the suitability of each applicant for employment:

False information from the applicant about prior employment.

An unexplained record of job hopping by the applicant.

Long unexplained periods of applicant unemployment.

False information from the applicant about his education, United States military service, or United States citizenship.

Dishonorable discharge of the applicant from United States military service.

The applicant has indulged in criminal, dishonest, or notoriously disgraceful conduct.

The applicant is a habitual, user of intoxicants to excess.

The applicant is currently addicted to the use of narcotics or controlled substances.

The applicant has a record of previously unsatisfactory employment or of reckless irresponsibility of a wanton nature indicating poor judgement or instability.

The applicant has a record of illness, including mental illness, that in the opinion of competent medical personnel would cause severe difficulty in

judgment or reliability, with due regard to the duration of such illness.
The applicant has a previous history of fighting or uncontrolled temper.
The applicant is subject to to coercion, influence, or pressure likely to produce
   actions contrary to company interests.
The applicant has a history of financial irresponsibility.
The applicant has a history of insubordination or disloyalty to previous
   employers.

The purpose of a background investigation, as distinguished
from technical screening, is to determine if the applicant has
been comprehensive and truthful in his answers and whether
past history demonstrates the presence in his background of any
relevant unfavorable information. The function of the employment or other department making the final hiring decision is
to be sure there is no information concerning either the applicant's technical qualification or past history that would require
his rejection as unsuitable under the standard.

The division of functions among technical, employment, and
investigative components also helps to sharpen each within its
own proper limits so that the reasons supporting each judgment
are evident. Also, if the exercise of the functions is likewise so
divided that no one person or position makes all the judgments,
there is a greater likelihood that the final decision will be objective and free from the influence of role conflict. This point will
be developed further in the later discussion of investigative
resources.

The Application Form—A Basic Tool

The application blank is the most important single document
an employment applicant will submit. It is generally the first
document he completes and that from which persons along the
processing route will draw information. It should be designed
to draw the maximum amount of lawful, relevant information
from the applicant and to prevent easy omission or misstatement.

The first entry, the applicant's name, is itself an item about which more information is often required than is asked for. An applicant who now calls himself Jones might have attended school or have been previously employed under a different name. Legal change of name, for whatever reasons the person involved may have, establishes a new legal identity. Casual change of name does not. In either case, the person will have created a history under both names. An employer has a right to know the relevant portions of that history regardless of the name the applicant was then using.

Some fair employment practice commissions have held that it is an improper question to ask if an employment applicant has changed his name. The obvious reason for this rule is that the name change may have been made to disguise nationality or ethnic character, neither of which is a permissible basis for employment discrimination. However, if the application form requires the applicant to indicate, in regard to specific past activities such as schooling or previous employment whether further information about any other names by which he was then known is needed to permit the prospective employer to make inquiry at the places and times involved, then the question is not asked for discriminatory purposes and is allowable.

The same principle guides the use of other questions that might be construed to penetrate areas protected by the law. In general, in marginal cases the employer will have to prove, not merely allege, that his purpose in asking the question is nondiscriminatory. Guides have been published by most states that have FEP legislation as to proper and improper application blank questions. Various subscription services cumulate these guides and other phases of employment practices laws and are very helpful in developing and updating the design of application blanks.[2] The following abbreviated checklist identifies the application form items that require careful attention to elicit the required information:

Checklist

Name. All names (including maiden name for women) necessary to check relevant past history.

Residence address. Not merely the current address but all previous United States addresses for some representative period such as ten years. That will permit continuity in investigation and will require the applicant to establish his whereabouts. Even when there is a detailed answer to the proper question, some unusual situations arise. In one case an applicant listed his activity as unemployed and his residence as Stormville, New York. An alert investigator recognized the community as the site of a state prison, in which, as it turned out, the applicant had been during the period. In listing residence addresses, the instruction should require chronological sequence to permit ready identification of gaps. Note also that the request only for U.S. addresses is to avoid appearing to solicit a foreign address, potentially a discriminatory request.

Education. The question should be carefully drawn to require specific year and month for (1) inclusive dates of attendance and (2) date of completion or termination and also to require (3) reason for termination, (4) degree or certificate or diploma awarded, and (5) the date the award was made. By listing years only, applicants have created the impression they possess degrees for which at the time of application they were still only candidates.

Employment. The following items are important for each employment that the applicant has had during a representative period such as 10 years: (1) name of the employer, (2) address of the employer, including street number, city, and state, (3) specific job title last held by the applicant, (4) compensation received by the applicant at time of termination broken down into salary and bonus or other incentive, (5) the date of termination, (6) the reason for termination, and (7) the name of the applicant's last immediate superior. With all these facts it should be possible, unless the former employer is no longer in business, to establish some data about the applicant. The application form should also require that periods of self- or unemployment be appropriately identified in the proper place in the chronology.

Licenses or special qualifications. Some applicants will hold special credentials such as a license as professional engineer, membership in a state bar, or license as CPA. The question on special qualifications should require the applicant to identify the license or certificate by its common name, the issuing authority and location, the date of issue, the certificate or license number, if any, and whether the license is currently valid. Verification of special licenses is sometimes difficult without one or another of those items, and different states index their records differently.

Military service. Military service in the U.S. armed forces is an appropriate topic for inquiry. To be certain that the service can be verified, it is almost always desirable to have the applicant produce a DD 214 (certificate of separation) or other equivalent official document if his service has been recent and he might be expected to possess the document. In some states, the certificate cannot be required until after employment. In every case, the information about service should include the branch, the inclusive dates, the serial, service, or Social Security number, the rank at time of separation, the character of the separation, and the last station or post of assignment. For recent separatees it may be possible to obtain verification at the last station, particularly if it is a regular post rather than a processing station. For earlier separatees, such as those who served in the Korean action or during World War II, records are available at the Federal Records Center.

Personal references. At least three references who are not related to the applicant should be requested. Their full names, residences, and business addresses should be required. Although a personal reference may be presumed to be favorably disposed toward an applicant who lists him, that is not always so. In any event, a personal reference may be able to provide a lead or clue to activities or whereabouts during a period about which the application is unclear or deliberately misleading.

Criminal or police record. A general question should be included in the form substantially as follows:

> Have you ever been convicted of or pleaded guilty or no contest to any serious crime, civilian or military? If the answer is yes, list the crime, the date, the police department or other agency involved, and the final disposition.

There are some current indicators of a trend toward narrowing the scope of such a question. The U.S. Civil Service Commission has removed the question from standard application forms, and some states, in illustrating lawful questions for application blanks, restrict the language to "convicted."

However, a potentially very significant decision was made by the U.S. District Court in California on July 28, 1970, in *Gregory v. Litton Systems, Inc.,* (316 F Supp. 401). It held a policy of rejecting employment applicants merely because of multiple arrests to be in violation of Title VII of the Federal Civil Rights Act of 1964 because it "has the foreseeable effect of denying black applicants an equal opportunity for employment." The key fact in the case was the very much larger number of blacks arrested than whites. The court held that, irrespective of intentions or equality of application of policy, the question was discriminatory. The defendant was enjoined from asking future applicants, or otherwise obtaining, information about arrests that did not result in convictions. No decision was rendered on the defendant's use of convictions as a basis for rejection.

This decision, as affirmed by the 9th circuit court, became the general law and requires all employers covered by the Civil Rights Act to discontinue the type of question that asks for arrests unless it can be shown that mere arrest record is not used as a basis for applicant rejection or discrimination.[7]

Names of relatives with the company. The names of relatives with the company can be particularly helpful in establishing connections and relationships that might otherwise not appear and that contain a potential for conflicts of interest. It is also necessary to implement a policy of not permitting related persons to work in the same direct line of authority, a not uncommon practice.

Authorizations and Agreements

In addition to the checklist questions and any other permissible questions related to the specific prospective employer's particular requirements, it is very useful to require the applicant, when he signs the application form, to execute an authorization and release regarding subsequent investigation of his statements and an agreement, if employed, to be bound by the employer's rules regarding security and safety. The authorization and release might follow this form:

> I hereby authorize [name of employer] or its agents to conduct such investigation of my application for employment as may be necessary in their sole discretion. I authorize all persons who may have information relevant to this investigation to disclose it to [name of company] or its agents and I release all persons from liability on account of any such disclosure.

Including only the authorization even without the release from liability would still be quite useful, because some records sources, colleges and universities for instance, are announcing policies against disclosing any information concerning a former student without the student's express consent. The applicant's signature under such an authorization on an employment application and a photocopy of the application for the school's files will often resolve both difficulty and the delay.

The legend and agreement regarding the employer's rules and regulations could read like this:

> I agree, while on [name of company] property in connection with this employment application, and at all times after employment if I am employed, to observe all [name of company] rules and regulations concerning security and safety.

Finally, there should be a caution to the applicant about the nature of the relationship created by his tendering the application and of the effect upon his application of false or misleading information furnished by him. The caution should appear close to the authorization release and agreement to observe security rules, and it should most certainly be on the same page and near the applicant's signature. It could read like this:

> I understand that my tendering or [name of company] accepting this application does not constitute employment or an offer of employment. I also understand that any false or misleading information furnished by me on this application form or in connection with the application for employment may result in the rejection of the application or, if employed, in the termination of employment.

Some states have statutes that make it a crime to obtain employment by submission of false information. If such a statute is available, it may be useful to add to the suggested caution about false or misleading information some reference to the law. A check of local statutes at the time applicant blank forms are designed will reveal whether such a law does exist.

Review of the Application

The most carefully designed employment application form will fail of its purpose if the applicant is permitted to complete it inadequately. Prior to its being accepted, a completed application form should be reviewed with the applicant by a responsible member of the employment staff or other competent function. If the questions have been properly drawn, it will not be possible for the applicant to use dashes, N/A, or other ambiguous notation to answer. Each prime question should require a yes or no reply. Related subordinate items can then be answered or skipped, as appropriate.

The key questions and items—those discussed in the checklist—should always be answered specifically and clearly in the applicant's own hand or in typescript if prepared by the applicant. Some firms follow the practice of interviewing the applicant and having the interviewer write the answers. That may be easier in some cases, but it leaves open to doubt whether a particular answer that is later the subject of inquiry or challenge was really given by the applicant. It is preferable, from the security point of view, for the interviewer to inspect and review the form and assure that all questions have been properly answered by the applicant. It is also helpful, after the applicant dates and signs the form, for the interviewer to witness the signature. That will establish the authenticity of the form in the face of any later challenge.

If the review of the application at the time it is completed reveals any questionable items, prompt follow-through by the interviewer can save hours of investigative labor and many dollars of cost. Such action will often permit final disposition of the applicant to be made at that point. In this regard it is interesting that the investigations reported in the survey discussed earlier[3] produced a noticeable change in the quality of original interviews by employment department personnel. The knowledge that mistakes or oversights made in the interview would later probably come to light through the investigation was an inducement to the interviewers to do a more thorough job. Even practiced employment interviewers will allow application forms to slip through in unsatisfactory condition at times. Periodic audit of the forms processed by the employment unit, independent of whatever additional insights are gained through investigations, will improve the quality of applicant processing.

Another important item that must be considered at the point in processing where the applicant has submitted and the company has accepted the application form is that of a hiring commitment. Some job skills are in scarce supply, and any undue delay in notifying acceptable applicants may cause them to look elsewhere. If the available investigative resources permit the investigation to be accomplished prior to final offer of employment, the job offer should await the results of the investigation. A later discussion of investigative time service will deal with elapsed time between application and investigative results. In those cases a particularly insistent applicant can be offered a contingent starting commitment. A certain date is suggested as his starting date contingent only upon the results of investigation.

Because all the technical and objective screening will have been finished, the only legitimate reason for the later investigation to upset the contingent hiring would be that unfavorable information was developed. An applicant who is unwilling to await such an outcome should be regarded with extreme skepticism. He must have believed the company fair and a good place to work when he decided to submit an application. If the only contingency now is verification of his own statements, he has no reason to believe the company will deliberately repudiate its contingent offer for any other reason than his veracity.

Period Preceding Decision

It is to the advantage of both parties to leave the offer in a contingent state. If, after investigation, the company does find reason to reject the application, it will not have incurred either

a legal or equitable duty to reimburse the applicant for any relo-
cation expenses. The applicant should be discouraged from in-
curring any transfer expense during the period.

Sometimes an applicant is apparently acceptable but requests
that his current employer not be contacted until after a firm
offer of employment has been made to spare him loss of his
present position. This demand also can be accommodated with
the contingent offer. It is not wise for a prospective employer
to make a hiring commitment while agreeing not to review what
may be the single most relevant aspect of the applicant's employment history. If the applicant understands that contact will
be made only after he has received a contingent hire offer and
that the same conditions apply to his statements about the
present employer as to all other statements on the application
form, both parties have a reasonable basis on which to proceed.

The practice of asking an applicant to indicate on the application form the employers whom the prospective new employer
should not or may not contact is not a good one from the employer's point of view, and he is under no obligation to continue
or adopt it. In cases of executive recruitment in which the
prospective employer makes the first overture the situation is
somewhat different. Care must always be exercised in those situations not to compromise the potential recruit. But with walk-in
applicants, their own conduct speaks for itself. If, in a given
case, some special problem exists, the applicant can discuss it
with the interviewer and some accommodation can be reached,
perhaps a form of contingent offer. A claim could be made out
against a prospective employer who gratuitously invited the applicant to deny access to the present employer if later the applicant were compromised through actions of the prospective employer and not hired by him.

The employment application relationship is generally one in
which the parties deal at arm's length. However, agreement by
the employer to observe a condition and later breach of that
condition, intentionally or through the employer's negligence,
with resulting damage to the applicant would create arguable
equitable grounds for redress.

When investigations cannot be completed prior to employment, effort should be made to assure some escape provision. Usually it is assured by reliance on a probationary period. The employer may terminate the employee at any time during the probationary period without liability. If during that period unfavorable investigative results are reported, they can be the basis for termination. The two problems with this approach are, first, that the cost of hiring (including the cost of investigations) has already been incurred and, second, that there may now be reluctance by operating supervision to lose an employee already in place, despite the unfavorable investigation. The longer the elapsed time between the hiring date and the investigative results, the more expensive and difficult will be the termination. The other possible consequences mentioned earlier should also be considered. An employee ultimately terminated because of an undesirable background may involve the employer in difficulty during the short period of employment precisely because of the features developed through the investigation.

It is in all employers' interests to weigh quite carefully any offhand assertion that preemployment investigative delay will cause loss of critical manpower. If a reasonable investigative time service basis can be established, actual experience on a selective trial should be accumulated before a policy decision is made. The rates of occurrence of unfavorable data in the survey population previously described indicate that even in the face of sharp demands to attract limited manpower, totally premature hiring may be an illusory benefit.

Particularly in cases involving collective bargaining contracts with provisions for just-cause discharge and grievance arbitration, the employer must assure himself that he has properly used the prehire or posthire probationary period to full advantage. There are cases in which even the applicant's palpable misstatements, discovered after the probationary period, have been held insufficient ground upon which to base a discharge subject to the just-cause requirement either because the potentially undesirable behavior has not materialized or the claimed but nonexistent qualification has not affected the employee's work or because the employer had time to discover the misstatement by diligent inquiry and failed to do so. In either situation an employer may be obliged to continue in employment a worker who is really marginal even if not so seriously defective or inefficient as to be severable for cause.

The Scope and Content of Investigations

Three considerations will affect the extent to which investigations are carried: cost, time available, and the investigative resource at hand. The cost of a background investigation should be regarded in the same way as any other cost component in the recruitment-and-hiring cycle. It should be held to the minimum consistent with objectives but not allowed to obscure the need for the investigation. Employers readily allocate sums for classified and display advertising in newspapers and trade media as a necessary part of the hiring cost. The amounts actually spent vary with profitability and cash availability but, except in the most austere situations, are not totally curtailed. Personnel departments are familiar with advertising and recruitment expenses and accept them fairly readily. Investigative expense, being less familiar, is sometimes cut without adequate reflection.

The absolute cost in dollars will depend most upon how much investigation is conducted and by whom. Cost ranges extend from $4 or $5 for a rather abbreviated report by a mercantile agency or credit investigating concern to $100 or more for a comprehensive report by a private investigating firm. The typical employer requires a good deal more information than is provided by the least expensive type and less than would be provided by the most expensive one. Arriving at the appropriate amount of investigation and allowable related cost is a task that calls for the combined capabilities of the personnel manager and the security manager. Skill in the use of available investigative resources can optimize the benefits from investigative expense. That is the reason why a security professional, either on staff or primarily interested in the company as a consultant, should assist in the design of investigative requirements. Knowledge of what resources there are and how to utilize them comes from experience and training in the field of investigations. It is not reasonable to expect a personnel manager to possess that knowledge if he has not had the experience or training that would supply it.

The scope of investigations suggested in the following paragraphs is based upon the results of thousands of investigations of the applicant type for both government and industrial employment. The lengths of time to be covered are based on actual experience with relevant, unfavorable information developed over such time spans.

Employment

Most unfavorable information about most applicants will come from or be developed through sources connected with their prior employment. A sufficient period of time should be covered to allow for that situation. For a production worker or shop and office entry level employee, a suggested minimum is the immediately preceding five years or last two employers, whichever is the longer. For personnel on a technical or supervisory level, the period should be extended to ten years or the last three employers, whichever is the longer. For management and executive personnel, the entire business career should be investigated for a minimum of 15 years.

The information of importance at each place of former employment includes (1) inclusive dates worked, (2) position title and compensation at termination, (3) nature of duties, (4) the reason for termination, (5) eligibility for reemployment, (6) reputation for honesty, diligence, and good conduct, and (7) prior employment as indicated in the records.

Education

For positions that do not require specific educational background, no investigation of education is necessary, but when specific education is required for the position or is a basis upon which wage rate is determined, education should be verified. College level and graduate degrees should generally be verified in all cases. For each school checked, the following items are significant: (1) inclusive dates of attendance, (2) program or course in which enrolled, (3) degree, diploma, or certificate awarded, (4) reputation for honesty and good conduct, (5) the reason for withdrawal if the withdrawal or termination was prior to graduation or other normal completion, (6) and, if specifically relevant to the hiring decision, wage rate, or job level, the class standing and average.

References

Interviews with references in situations in which all other items are regular and no gaps or unexplained periods exist are generally not useful. When a question of whereabouts or activity during a given period arises, a reference can often supply the data.

Reputation at Residence

Although they are sometimes among the most difficult to accomplish, inquiries at present and former residence neighborhoods can occasionally confirm unfavorable information regarding alcoholism, instability, and histories of illness or accident. Generally, unless other indications are developed through the employment checks and unless the investigative resources are thoroughly professional, neighborhood inquiries can be eliminated. When they are conducted, coverage of the most recent two years or the specific period of past time in question should be adequate.

Police and Credit Records

Some police department used to make criminal record information available to a bona fide employer at no cost; others charged a fee for processing a request. Still others, principally the metropolitan departments, did not release any criminal information except to the applicant or at his request. Since adoption of 28 CFR 20, et. seq., most police departments will make criminal history record information available only to those persons authorized by federal or state statute or executive order to have it. Most employers do not qualify under these criteria. Unless some aspect of the investigations suggests the existence of a police record, effort to determine whether there is one can be eliminated. When evidence is unavailable but there is solid reason to suspect that a record exists, there are other sources. The dockets in various criminal courts are public records, and if the applicant was tried and the court and date of trial can be determined, the record will be available.

Credit bureau information is useful with respect to histories of trade account delinquencies, suits or judgments, and personal or business bankruptcies. Care must be exercised because such records are often partial or preliminary and should really be only leads for further check. Fairness requires that old or incomplete information not be used in forming a judgment of an applicant. Such information is so frequently found in credit bureaus that consumer protection statutes are being enacted to limit its availability and to increase the bureau's liability for inaccuracy in reporting it. Some such statutes do not include a prospective employer as one allowed to order a credit report unless the order is authorized in writing by the subject of the inquiry.[4] If other information suggests financial instability or credit problems, the local credit bureau can be checked (for a fee and subject to whatever statutory limitations may exist). The results of that check should be updated by further inquiry with the creditor or other prime source. In otherwise routine investigations, credit bureaus and agencies can be eliminated.

In addition to various state laws dealing with credit reporting agencies, Congress passed a comprehensive measure, on October 26, 1970, as Title VI of the Consumer Credit Protection Act. Now codified as 15 U.S. Code 1681, et seq., the provisions became effective on April 25, 1971. Every enterprise that uses any kind of commercial agency, mercantile reporting bureau, or private investigative organization for any information regarding employment applicants or candidates for promotion or reassignment should be thoroughly familiar with this new section of the U.S. Code.

The section imposes the obligation on any user of commercial investigative reports in connection with hire, promotion, or other "employment purpose" that he notify the candidate (called the consumer in the act) *in writing* that an investigative report may be utilized and, if thereafter requested by the consumer, that the user also notify the consumer of the kinds of information he has requested in the investigative report. Finally, if a decision unfavorable to the consumer is made as a result, in whole or in part, of the consumer investigative report, the user must so inform the consumer and must advise him of the identity of the consumer report agency used. The Act then gives the consumer certain rights against the agency in respect to the information contained in its files and in the report in question. Because there are both criminal and civil penalties in connection with the willful or negligent violation of the Act, all employers should be sure that they are in compliance with the provisions of the Act if they use or plan to use outside investigative resources in screening employment applicants.

Extensions of Investigation

When preliminary results indicate unfavorable information by any criterion used for selection, further investigation should be conducted to resolve the matter and ascertain its full scope. That does not always involve continued investigation with outside sources, but it may require another interview of the applicant. For example, if an executive under consideration claims to have been paid $25,000 per year at his last place of employment but the investigation is either unable to determine the amount or develops some significantly smaller amount, the applicant himself may be asked to volunteer some confirmation. The simplest would be his form 1040 file copy for that year. Another illustration would be claimed graduation with a degree from a college where no record can be found. The applicant can be asked to produce the diploma or grade transcript or even the graduation bulletin with his name in it.

When all reasonable efforts have been made to resolve a serious doubt and the doubt cannot be resolved, the employer should have determined, as a matter of policy, what to do with that applicant. The recommended policy is to terminate consideration. In given cases, for adequate reason, executive management can waive the policy, but that should require the matter to be brought to the attention of a member of executive management. It is argued by some that elimination of a candidate because of a doubt is unfair or even illegal. An employment applicant is not accused of crime, and no presumption operates in his favor. When a business organization employs a candidate about whom reasonable, unresolved doubt of suitability exists, a serious risk of potential harm is created through the vulnerabilities due to such a person's presence. If a decision to employ is made, it should be made as an exception, not a norm, and it should be made by a responsible senior official.

Investigative Resources

Who will conduct the investigations is often the first question asked in respect to applicant screening. It has been presented here as the fourth consideration because the determination of resource should follow, not precede, the other determinations.

There are three investigative resources. One is the customary reference check in which the prospective employer, through letters and telephone calls, contacts sources of information. The second is the use of outside agencies through which a commercial investigator is retained and works from the application form data secured by the employer. In those cases the agency is sometimes permitted to identify the actual employer interested and sometimes is not. The third resource is the use of professional investigative personnel on the security staff of the prospective employer. Many of the larger United States companies, particularly those identified with the defense industry, have used staff investigators regularly since the period of World War II.

Reference Check

The most common approach is the reference check. If the labor pool is from an entirely local, nonmetropolitan labor market and the employer is well known and respected, a satisfactory result is often obtained. However, when applicants come from a wide area or are recruited nationally and internationally and when the assigned personnel staff consists of inexperienced interviewers or clerks, the results are usually not reliable. In the first place, routine inquiries are given routine handling. Most personnel departments tend to deemphasize or omit really unfavorable data when they reply to reference check inquiries. They do so partly from the fear of making an actionable statement in writing and partly for reason of the sheer labor of constructing an appropriate reply to an inquiry that involves serious derogatory data. Another problem with reference check procedure is that replies, when they are received, are not properly evaluated and that outstanding requests are not followed up. Failure to receive a reply to a letter of inquiry is a caution signal. However, after a certain elapsed time, many employment departments close the file without having received anything. If no other resource but the employment department is available for investigations, the following minimum standards should be established:

Reply time. Outstanding inquiries on which no reply has been received will be followed up, by telephone, within a reasonable period, not more than ten elapsed days.

Ambiguous replies. Vague or ambiguous replies will be reviewed with the source by telephone for clarification.

File review. A responsible member of the personnel or employment department will review all closed investigative files to assure completeness and to make a determination about the character of the information contained in them.

Unfavorable cases. The unfavorable cases will be indexed in some way to assure that the same applicant will not inadvertently be accepted for employment at a later time.

Pending cases. When preemployment investigation is the policy or when postemployment investigation is related to a probationary period, a tickler file will be maintained to assure timely action.

Commercial Agencies

The second resource, the commercial agency, offers slightly more than the reference check, provided that adequate controls are maintained over the quality and timeliness of results. If a commercial investigative agency is utilized, it is helpful if that agency is permitted to reveal the reason for the inquiry. Potential sources of information will cooperate more readily when they know the identity of the real party in interest. Careful use of the authorization and release, and determination that the agency used is adequately covered by the appropriate kind of liability insurance applicable to any loss claim made against the employer, should permit broad discretion on this count.

When a commercial agency is retained, the information requirements developed by the employer should be discussed in detail. Often the commercial report that is provided exceeds the requirements of the client or does not cover a vital item. The report should be tailored to the specification to the greatest extent possible to conserve costs and save time.

Some agencies attempt to conceal the identities of the sources and informants used by them in compiling the report. That leaves the employer in the difficult position of having to rely on the agency not only for the investigative coverage but also for the reliability of the result. The choice is not a wise one. A reputable agency that is operating according to law and good practice should not be concerned about revealing its sources of information to a client with whom it has established a relationship. Some cases will be exceptions because the nature of the information or the sensitivity of the informant is extraordinary. As a rule, however, the agency used should be willing to identify its sources of information. The more costly reports by investigating agencies generally do identify the sources. The less expensive reports available from credit bureaus and mercantile agencies often do not. That should be a factor in final selection.

Restrictions on commercial agencies. Under Title VI of the Consumer Credit Protection Act, discussed previously, organizations that make consumer investigative reports are obliged to maintain comprehensive records of the identities of all persons (users) who order reports, of the information contained in such reports, and of the sources of such information. Restrictions are imposed on reporting organizations not to report many types of unfavorable information seven years old or more at the time of the report. Repetitive reporting of unfavorable information initially developed in the preparation of a consumer investigative report is also prohibited unless, in any subsequent reports, the matter is verified or reconfirmed by current investigation or unless the subsequent report is submitted not later than three months following the initial report. In some cases the provisions of this law may make significant changes in the scope of information previously obtained from mercantile credit and investigative agencies.

Personal visits by investigators. Another point sometimes raised both pro and con the quality of an investigation relates to whether the investigator personally visits the informants. The so-called street investigation is considered by some a more valuable effort than telephone or other impersonal communication. That depends almost entirely upon the skill of the investigator and the presence he shows or interest he is allowed to state he represents. To a trained, professional investigator with three or more years of experience in an efficient intelligence or law enforcement agency, personal contact is not necessary to good results. To an untrained learner who is gathering his experience while conducting your investigation, the contact may be necessary. Companies that use untrained recruits—generally the larger credit and inspection firms—have little choice but to insist on that approach to maintain their own quality standards. When the choice is from seasoned investigative resources, however, the point is not central. Proper audit on the reports of investigation by the client will determine whether the quality standards are being met. The technique can be left to the reputable and experienced agency.

Staff Investigator

The last resource—the in-house or staff investigator—has much to recommend it. A competent, professionally trained investigator can be relied on to accomplish each month from 30 to 40 investigations of the kind discussed in this chapter. The smaller the scope the greater the number of cases. Under stress conditions as many as 55 investigations can be accomplished.

Time and distance are not barriers, because the trained professional can make proper use of mail, wire, and telephone. Moreover, being on staff, he is under the quality control of the employer and is responsive only to the employer's requirements.
Finally, when they are not applied to applicant investigations or
when a priority interruption is proper, such personnel can attend
to other investigative matters within the facility such as fraud
and theft loss, gambling, and administrative inquiries in which
the skill and discretion of a professional are required.

As a matter of cost estimating, the survey data reported
earlier established that a continuing, average applicant case load
of 40 cases per month will support a staff investigator as the
fastest, most efficient, and most cost effective resource. As case
load increases, the staff increases in proportion. The most valid
way to make cost comparisons when using staff personnel is to
allocate all costs, direct and support, of their activity over some
representative period of time and number of cases and divide
the total cases completed into the total cost. The per case or
unit cost derived can then be compared with similar costs that
would have been incurred if a mercantile or other commercial
agency had been used.

It is always difficult to make a reliable cost estimate to fit all
situations because so many variables affect it. As a general guide,
however, the unit cost for investigations generally meeting the
standards here discussed, when done by a staff investigator, and
considering 1981 as the base period, should be between $100 and
$125. Commercial agencies using seasoned personnel will charge
between $75 and $200 for comparable results. Large mercantile or
credit reporting firms will offer services from a low of $5 to $10 to a
high of over $100. Reliability is related to cost. In the middle ranges
where all three resources are comparably priced, comparable results
can be expected. At the lower end the reliability will fall off con-
siderably.

Investigative Time Service

Timely results are as important as quality results. What is
a reasonable time frame within which to expect performance?

In part, that depends upon the resource. Outside agencies of either the credit bureau or private investigative type will require from three days to three weeks, or more, from receipt of inquiry. When many different offices of a commercial agency are involved in the inquiry because the applicant's background is spread over many locations, the elapsed time will be greater. A staff investigator should be able to complete most cases within an elapsed time of five working days from receipt of the completed application form to preparation of the results.

An added advantage of staff investigators is that in favorable cases no elaborate report need be prepared. The investigator's notes, suitably reviewed, can constitute the file and the results can be communicated to the employment department immediately upon completion of the field inquiry. Any time requirement beyond three weeks should be questioned. An investigation that requires that long is not timely. In such cases applicants may be lost or contingent offers become the rule or postemployment investigation the policy. There is also a kind of Parkinson's law involved. With the passage of time, investigative reports from commercial or private agencies will tend to take as long to complete as the client is willing to wait. If outside resources are used, a responsible employee of the client employer should maintain close liaison and monitor suspense dates with the agency.

The Hiring Decision

The last important question is who should make the hiring decision. The best answer is the employment department or the line manager, whoever traditionally has had the responsibility. It is not good practice for the investigator to make the decision because, as an investigator, his function is to gather the information, particularly the unfavorable information. Moreover, if the employment manager investigates or the investigator hires, there is a role conflict. One person cannot easily be the recruiter and the investigator simultaneously. Of course, when outside agencies are used, it must be some internal function that decides. But even then a staff security manager or director, not equipped with staff investigators, may use an outside agency to do the work and then play a role in the decision. If the standards and criteria are clear, and if the report is adequate (the job of the security organization or investigator) the decision to hire or reject should be evident to the employment department. In arguable cases a common superior or other senior can mediate.

Periodically, the hire-reject performance of the employment department should be compared with the favorable-unfavorable content of investigations to assure that policy is being carried out in both the screening and hiring functions.

REFERENCES

1. "The Case for Applicant Investigations," *Industrial Security,* October 1966, pp. 3ff.

2. An example is Employment Practices, a subscription service offered by the Commerce Clearing House, Chicago, Ill.

3. Note 1, supra.

4. New York State General Business Law, Secs. 380 et seq. last ammended in 1977.

5. It should be noted that increasingly statutes are prohibiting or restricting physical disability as a permissable basis for rejecting employment applicants. In this regard see, *Rehabilitation Act of 1973* (P.L. 93-112) and equal employment opportunity laws in most states.

6. Careful study is suggested of *Personal Privacy in an Information Society,* the report of the U.S. Privacy Protection Study Commission, published July 1977, especially Chapter 6—The Employment Relationship. This report has already prompted many employers to change employment policies and it may spur legislative action at the federal and state levels. The Commission made thirty four specific recommendations, some of which deal directly with pre-employment screening activities.

7. Gregory v Litton Systems, Inc., 472 F2 631.

Self Test Questions

1. What is the reason for conducting a background check of an applicant for employment?

2. What criteria would be appropriate in deciding whether information about an applicant should be considered unfavorable?

3. What would be the minimum information which should be required on an applicant questionnaire?

4. What extent of investigative coverage would likely provide a good insight?

5. What three resources can be utilized for background investigations?

6. What might be the cost-per-typical-applicant investigation for a high quality investigative agency and for an employed investigator?

7. What employment practice could assure a definite offer to the applicant but preserve the investigative option if an urgent response or reply were necessary?

8. Who should normally make the hiring decision after investigative results are completed?

= 12 =

Effects of
Changing Social Environment
on Security Planning

IN this chapter an attempt will be made to analyze current trends and tendencies in major social and economic forces. The attempt is relevant because the United States is passing through a period more significant than any since the post–Civil War reconstruction in respect to its impact on law enforcement and protection of private property rights. Fundamental changes have occurred and are still occurring in the way people see themselves and each other. For large subgroups the "establishment" is not what we are all trying to build and preserve, but the arch enemy. For the first time in many years there is widespread concern not only that people should be free from all public restraint in their private lives and personal occupations but that they should also be much less regulated in their public behavior. There are many influences at work to "liberalize" and "humanize" life in this country, although the exact meaning of those terms is by no means generally agreed on.

Major Influences for Change

The prime areas of concern to this study are the legal framework, the social structure, and the level of technology. Within each of these main areas are flowing currents and cross currents that have both an immediate and a long-range meaning for the manager planning a security program. Some traditional practices will have to be modified and some new concepts and techniques will be required to maintain acceptable levels of security protection.

Legal Framework

Revolutionary changes—to use the word in its nonviolent context—have taken place in the law over the last twenty years, particularly the criminal law. Our jurisprudence is fundamentally different from what it was in the post–World War II and Korean emergency years and even different in important ways from what it was during the early 1960s. We are shifting from primary concern with property and orderly process regarding property to primary concern with personal liberty and the conditions of individual human beings in specific situations. The change is concentrated in the new attitudes toward constitutional due process, that amalgam of rights, privileges, and procedures chiefly related to and regulated by the first, fifth, and fourteenth amendments to the U.S. Constitution.

Police may no longer accept even the entirely voluntary statement of a suspected person in custody without first warning the suspect that he need not speak, that, if he does speak, what he says may be used against him, that he may request and have a lawyer present, and that if he wants a lawyer but can't afford one, one will be provided for him. It is a truism in police circles that most crimes would not be solved without admissions or confessions from the criminals. The new restraints on questioning[1] will have a major impact on crime solution and police behavior. In fact, they already have had.

Strict limitations now apply to all the states on the extent to which a police officer may search a place or person and what he may seize during the search. The fourth amendment has always imposed those limitations on federal officers. As the result of recent decisions by the Supreme Court,[2] no court, state or federal, may permit the admission of materials seized or evidence acquired through an unreasonable search.

Police may no longer disperse groups merely by invoking statutes making it a crime or offense not to move on after being requested to do so. Persons lawfully present cannot be compelled to move merely because other citizens or the police don't like their presence.[3]

Public officials will be less inclined to suppress remarks or writings, publicly displayed, on the ground that they offend public decency or are lewd or lascivious. Since the obscenity decisions of the Supreme Court[4] have left many unanswered questions about what constitutes pornography or obscene material, many enforcement departments are maintaining a hands-off policy on that problem.

No American should regret these constitutional developments. Personal liberty is too valuable a right to permit its incidental or consequent erosion in the name of public safety. There has been a great deal of hand-wringing by some who maintain that increased personal liberty has cost too much in police efficiency and resulted in sharply increased crime. There was even some sentiment for impeaching former Chief Justice Earl Warren after the Miranda decision. Police efficiency is not a constitutionally mandated or guaranteed process. The privilege against compulsory self-incrimination is. Whatever may be expended on additional training and resourcing of police organizations to deal with escalating crime without violating basic rights, the price is far less than even limited sacrifice of personal freedom.

There are two points here of great importance for the enterprise manager. First, the police will not be available for protection of property in quite the same way they used to be. Enforcement of the criminal law will demand much closer cooperation of the business community with the police. Cooperation must include the willingness to sign complaining affidavits, to appear as prosecution witnesses, and to report to the police the situations that warrant police action. For many businesses that will be a rather marked change.

The second point touches on financial support of police. Public money to finance police is raised through taxation. Businesses will be expected to share a growing tax burden as the police needs for improved resources are translated into budget requirements. In addition, direct support should be considered through grants or donations to local departments for the improvement of training facilities, scholarship aid to officers who are continuing their formal education, and the establishment of sound community relations programs.

The policeman has an extraordinarily difficult task today. Basically the guardian of the commonwealth against crime, he must enforce the criminal law. But the law is in flux. Some of the community seek to restore its former orientation toward property and to enforce it vigorously against all "antiestablishment" forces. Other members of the community are champions of the humanizing trend. Sensitizing the policeman to respond properly in this delicately balanced situation involves extensive human relations training. Training requires resources. Resources require money. Business firms and the business community at large have financial capability to support extended police efforts to meet training and development needs. Community Chest and United Fund efforts are supported by business, mainly through financial aid but often also through personal efforts of business managers. Improved police training in community relations should be supported the same way.

Specific Legislation

Beyond the constitutional change of emphasis, there are statutes being passed in many states that will have an impact on the safeguarding of business assets. One employee-screening tool that was fast acquiring popularity—the polygraph or lie detector—has been denied to employers by law in at least nineteen states,[5] and others are considering the same kind of law. An employer is now prohibited from terminating or discharging an employee because of a single wage garnishment against him. In

Chapter 11 mention was made of new laws that restrict the availability of credit bureau reports to employers as screening devices. Other laws can be expected to limit the kinds of personal data an employer will be permitted to elicit from employment applicants. Background investigations will be more narrowly limited. Decisions not to hire because of unfavorable investigative reports now oblige the employer to inform the applicant of that fact. It is even likely that employers may be denied the right to probe past criminal records, at least directly by questionnaire. The removal of arrest questions from standard form applications of the U.S. Civil Service Commission and the development of a "Privacy Personnel Security Questionnaire" by the Department of Defense portend efforts to translate these voluntary government actions into counterpart requirements on the business community.

Whether or not such developments are for the common good, they will require modifications of practices. Alert business managers will follow the trends closely and keep their own protection programs in conformance with changed needs. For example, the simultaneous tendencies to require businesses to offer more employment to persons previously considered unemployable or undesirable and to restrict more closely the background inquiries made about such persons put new pressure on the hiring process. Business managers must be aware of their position and must bring their arguments and requirements to the public debate.

Collective Bargaining

Two claims are often made for the arbitration of grievances arising under collective bargaining contracts. One is that arbitration, not being bound by legal rules of evidence, can deal with the obvious equities in a prompt and forthright manner. The other is that an arbitrator is not bound by prior decisions of other arbitrators. There is no doctrine of *stare decisis* that gives precedent weight to earlier awards. Actually, many arbitrators insist on all the procedural and substantive safeguards of legal due process when hearing disciplinary grievances, particularly those involving discharge or long suspension. It is also common

practice, at least within the framework of a single contract between one union and one employer, to agree that decisions will
have precedent value.

The insistence upon criminal due process is particularly interesting in the light of changes in the actual criminal procedures.
There are many arbitration awards that discuss discharge as the
"industrial equivalent of capital punishment" and that insist on
safeguards to protect the terminated employee very much in
the way the accused capital offender is protected. The fact that
so many arbitrators, particularly labor arbitrators, are lawyers
or come from legal backgrounds may explain the insistence.

The net result often is that a discharged employee is allowed
by the arbitrator to claim equivalents to the privilege against
compulsory self-incrimination and the right to be secure against
unreasonable searches. Cases in which accused employees have
furnished written statements admitting the acts charged have
been determined on the voluntariness of the statement. These
are essentially applications of due process. With the changes
brought by the Miranda and following decisions, it may be
anticipated that arbitrators will come to expect warnings for industrial employees before their voluntary admissions or confessions will be heard.[6]

The point to be noted is that business managers who are
preparing discipline cases involving security issues such as theft
or gambling should be sure that they follow closely the developing due process requirements for criminal procedure. Regardless
of whether or not the application of such principles to industrial
discipline cases is an appropriate exercise of the arbitrator's
authority, the application is being made and will probably follow
closely future changes in due process concepts. In short, when
a case against an employee that may be determined ultimately
in an arbitration is being prepared, it should be handled as
though it were subject to all the procedural requirements of an
actual criminal prosecution. Because the security organization
plays a major role in the development or investigation of many
discipline cases, it should be aware of and responsive to such
requirements. It is the responsibility of the enterprise management to insist that it be.

The Social Structure

Work is now regarded as a significant rehabilitative factor for persons who are socially maladjusted. That includes former prisoners, the culturally or economically marginal with few or no industrial skills, and those suffering from various forms of alienation. Work and workplace relationships are believed extremely important for the health and productivity of those very persons whom the business community has traditionally refused to employ.

The new attitudes will require businesses to respond in voluntary ways, or legislation will be enacted to prohibit discrimination because of character, reputation, former life style, or attitude and temperament. An employer today may discriminate against particular applicants for any of those reasons. The legislative approach in preventing discrimination on the grounds thus far embraced—that is, ethnic origin, religion, age, sex, and participation in labor activities—has generally involved prohibitions against asking applicants any questions on those subjects or otherwise seeking to acquire such information prior to employment. It is not unlikely that the same approach would be followed for any new grounds included in the statutes. Employers might be foreclosed at some future time from asking about prior arrests or convictions, about former mental illnesses, even about disciplinary infractions at places of former employment. If that were so, the security vulnerability inherent in unsuitable or unreliable employees would be greatly enlarged.

Rather than wait for legislation, the business manager should prepare now by adapting his employment screening practices. If an applicant with an unfavorable history but with an apparent sincere desire to modify his future behavior is employed despite his background (the actual practice in many "opportunity" programs now being introduced) the employer should at least be aware of the potential problem and be prepared to deal promptly with recurrences.

Controlled experiments should be encouraged for employment of the problem case. The controls, of course, are thorough investigation and evaluation of the past history, counseling at the time of hire and periodically thereafter, and prompt attention to facts that suggest recurrence of the undesirable behavior in the new job. This means an increased utilization of preemployment screening while that resource is still available. Businesses now may choose not to investigate, but they still can refuse to employ. Expansion of the laws may deny them the right to investigate and yet compel them to employ in certain cases. The trend is apparent.

Other developments indicate further changes in traditional standards. A decade ago any sexual behavior except conventional sexual relations between an adult male and female, generally husband and wife, was criminal. Homosexuality was not only a crime; it was a specific ground for refusal to grant government security clearances and a regular ground for refusal to employ or to terminate employment in industry. More recent attitudes toward consensual deviant sexual conduct of adults reflects increasing conviction that it should not be considered criminal. It may be that in the future there will be more sentiment expressed in favor of softening laws in this area and withdrawing the subject from public regulation. Employment of discovered or admitted sexual deviants will be harder to terminate summarily.

Another major social force operating to change traditional practices is the highly developed sense of individuality and personal dignity on the part of young people entering the labor market. The specter of a depression and national enthusiasm for a major "approved" war have, in the past, conditioned job applicants to accept any work simply for the security of employment or to accept any screening and disciplinary requirements for the sake of "national defense." Today, new, young workers question every restraint and every attempt to extend the influence of the workplace into their private lives away from the job. The attitude is in part a result and in part a cause of a more general preoccupation with privacy.

Committees and individual members of both houses of the Congress and of many state legislatures have made frequent statements about the invasions of privacy being committed in the United States. From collecting and central computer processing government data about citizens' backgrounds to handling personal information on job, credit, and other applications, contemporary information management is cited as an imminent if not real and present danger to society. The arguments run both to denying central facilities the right to process such information and to denying an initial input activity the right to acquire it in the first place. There is stiff and growing resistance to anyone, government or private, inquiring into the private lives of citizens. This is coupled with and reinforced by the feeling that individual performance in a job or activity is the only valid criterion of one's right to hold it. Antecedent information and historical data about performance on other jobs and at other times is regarded as irrelevant.

The Level of Technology

One result of automation is generally to change industrial processes from labor intensive to capital intensive. Human workers are replaced or augmented by machines and devices. That introduces new physical assets to the workplace and creates a requirement for different kinds of security measures. In large industrial complexes with few people, emergency responses themselves must be automated. Plant fire brigades must be replaced by fixed, automatic detection and extinguishment equipment, as one example.

New technology also makes some conventional skills obsolete while simultaneously creating new crafts and occupations. The newest, those relating to data management and computer operations, are introducing greatly enlarged security vulnerabilities to thefts of money and information (discussed at length in Chapters 6 and 7), but they are, at the same time, providing personnel to assist in the development and management of computer-controlled security systems (discussed in Chapter 10). The modern manager who is concerned with protection of his enterprise's assets will recognize both effects of technological change and will be as quick to benefit from the one as he is to avoid being the victim of the other.

Because of the increasingly intimate relationships between familiar industrial activities and the more esoteric areas of cybernetics, operations research, and control of huge information systems, much of today's typical security program is incapable of being responsive to the prospective new needs. New types of personnel will be required to discharge security responsibilities. Their skills and training will have to be in the new technology as well as in law enforcement. Integrated security systems of tomorrow will be as much applications of communications engineering, telemetry, and microelectronics as they will be of theories of crime prevention, patrol, and detection.

Predictable Changes

The changes and modifications in security practice that may be required because of the changing legal, social, and technical environment can be summarized as follows:

More prevention and less detection will place greater emphasis on passive security countermeasures such as sophisticated alarm and access control systems than on investigation and disciplinary treatment of offenses.

In the discipline situations that do arise there will be *greater need for solid cases against employees charged with misconduct,* which will require proving responsibility beyond a reasonable doubt (the criminal rule) rather than merely by a preponderance of evidence (the civil court criterion). It will also mean that the case must be built with due regard for enlarged due process requirements formulated in the area of criminal justice and applied to the industrial scene. One of the new principles may well be the right of any collective bargaining unit employee being disciplined within the context of a union contract to insist on union representation not only at the formal grievance hearings and arbitration but at every stage in the discipline. That would be an analogous application of the right to counsel, one of the major due process requirements coming from the Miranda decision.

New and different security personnel will be required; they will have to understand systems operations of various kinds and be able to function rapidly and at high orders of complexity. The requirement will be much more for intellectual ability than for physical condition or prowess.[7]

Previously unacceptable personnel will be hired into the work force; there will be those with criminal records and those with records of violent political activism. Sexual behavior will no longer be an allowable basis for candidate rejection. Prior personal history may be protected against employer inquiry. These changes will require even higher levels of confidence in passive security controls systems.

Preparatory Steps

To move with the changes and take maximum advantage of new opportunities for improving security programs, enterprise managers will first of all insist upon senior security personnel who are more management generalists than their predecessors have been. That will involve the acquisition by security professionals of skills in information management, engineering applications, and systems operations. The management development target for security personnel will be the function of total-risk manager rather than a law enforcement or crime prevention specialist. Prevention, control, and indemnification are the prime components of effective risk management. The security manager must become increasingly skilled in the operations of all those components.

Human relations capability must be extended. Security personnel, somewhat like police personnel, operate at the social fracture line. They are at the same time and for the same actions friends to one segment of the community and enemies to another. They must understand the ambivalence of their role and not believe they are expected to "win" in each confrontation situation.

Senior management should make room in its ranks for the risk manager and expect of him the same degree of skill and contribution as from other senior managers. The sheer size of its total security vulnerability will require the business enterprise to establish effective security countermeasures and to staff the security management function with personnel of broad competence.

The sources of competent manpower will not be the traditional ones. Technical exposure and overall management capabilities are not generally acquired in the law enforcement or investigative agency. Businesses should move now to develop effective career ladders in their own organizations for the preparation of tomorrow's security or risk managers. Realistic entry level requirements, balanced among formal education, business or commercial experience, and specific loss control training, are needed. Employing an ex-agent or ex-policeman to develop and manage what is essentially a business activity will no longer be the best appoach. There will, of course, still be transition opportunities from public service to business, and there will be employment situations in which the ideal candidate will be one with recent and extensive law enforcement exposure. But the direction and control of industrial protection programs, with increasing frequency will be placed in the hands of well-rounded business managers with diversified skills.

Finally, the serious enterprise manager will develop quantitative approaches to his security needs. He will find ways to measure the cost of probable security losses and will budget adequate resources to neutralize them. The prevention and control elements of risk management will receive as much attention as the indemnification element. The related functions of fire prevention and extinguishment, plant protection, casualty insurance administration, investigations, and information security will be combined into a unified management function.

Conclusion

If the predictions in this chapter have not dealt with every phase of industrial security, they have tried to see the main ones. There will be other changes that, for the most part, will be reflective of changes in the total community. By preparing for the significant changes, concerned managers will be in a good position to respond to incidental changes as they too occur.

In the short space of time between World War II and the Vietnam conflict (so short that it seems we will continue to reckon history as "times between wars") the nature and extent of risks of loss to industrial property have changed basically. The signs of further change are apparent. The emergence of conglomerate enterprises is combining the already enlarged loss risks into greater concentrations. It is also enabling serious organizations to acquire high-level professional competence in the special function of industrial security and to organize effective, systematic programs of prevention and control. Attention to these new dimensions of the old problem should be paid at the very highest rung of management. With informed policy decisions from that level, it will be possible to control security losses despite the wide variety of new forms they have already begun to take.

REFERENCES

1. *Miranda v. Arizona,* 384 U.S. 436.

2. *Mapp v. Ohio,* 367 U.S. 643.

3. *Shuttlesworth v. Birmingham,* 382 U.S. 87; *Brown v. Louisiana,* 383 U.S. 131.

4. *Ginzburg v. United States,* 383 U.S. 463; *Stanley v. Georgia,* 394 U.S. 557.

5. Alaska, California, Connecticut, Delaware, District of Columbia, Hawaii, Idaho, Maine, Maryland, Massachusetts, Michigan, Minnesota, Montana, New Jersey, Oregon, Pennsylviania, Rhode Island, Washington and Wisconsin. In addition, the states of New York and Pennsylvania specifically prohibit use of the P.S.E. (Psychological Stress Evaluation).

6. Interrogation without union representation for a bargaining unit employee who has properly requested such representation is an Unfair Labor Practice. (N.L.R.B. v. Weingarten, 95 SCt. 959).

7. See requirements for Professional Certification of American Society for Industrial Security, e.g.

Self Test Questions

1. What significant changes have occurred in the legal framework of security over the past 20 years?

2. May the polygraph be used as a condition of employment in all states in the U.S. today?

3. What is the significance to security investigations of N.L.R.B. v. Weingarten?

4. What further changes in the social and technical environment are likely to affect security in the near future?

Index